Opioid Peptides

Volume I

Research Methods

Authors

József I. Székely, M.D., Ph.D.
Deputy Head, Medical Department
Institute for Drug Research
Budapest, Hungary

András Z. Rónai, M.D.
Staff Member
Institute of Experimental Medicine
Hungarian Academy of Sciences
Budapest, Hungary

CRC Press, Inc.
Boca Raton, Florida

Library of Congress Cataloging in Publication Data

Szekely, Jozsef I. (Jozsef Ivan)
  Opioid peptides.

  Includes Bibliographies and indexes.
  Contents: v. 1. Research methods --
v. 2. Pharmacology -- v. 3. Opiate receptors
and their ligands.
  1. Endorphins. I. Ronai, Andras Z.
II. Title. [DNLM; 1. Endorphins. QU 68 S9970]
QP552..E53S93   1983   615'.78   82-12860
ISBN 0-8493-6235-0 (v. 1)
ISBN 0-8493-6236-9 (v. 2)
ISBN 0-8493-6237-7 (v. 3)

This book represents information obtained from authentic and highly regarded sources. Reprinted material is quoted with permission, and sources are indicated. A wide variety of references are listed. Every reasonable effort has been made to give reliable data and information, but the author and the publisher cannot assume responsibility for the validity of all materials or for the consequences of their use.

All rights reserved. This book, or any parts thereof, may not be reproduced in any form without written consent from the publisher.

Direct all inquiries to CRC Press, Inc., 2000 Corporate Blvd., N.W., Boca Raton, Florida, 33431.

© 1982 by CRC Press, Inc.

International Standard Book Number 0-8493-6235-0 (Vol. 1)
International Standard Book Number 0-8493-6236-9 (Vol. 2)
International Standard Book Number 0-8493-6237-7 (Vol. 3)

Library of Congress Card Number 82-12860
Printed in the United States

FOREWORD

The neurosciences are among the most rapidly and spectacularly expanding branches of contemporary knowledge. This most impressing spreading of these important areas of sciences in the last two decades is documented by a series of important discoveries whereupon our general picture on the organization and on the functions of the brain is dramatically changing year by year. No wonder that on the occasion of the last International Congress of physiological sciences, organized 2 years ago this time by our scientific community here in Budapest, Hungary, about one third of all papers presented dealt with nervous regulatory phenomena.

The sudden expansion of our knowledge of the activities of the nervous system is due to new, mostly unexpected findings among which the topic of the present monograph is a typical example: two decades ago nobody could even guess the bursting career of the opioid peptides.

The authors of this book were fortunate to join the research endeavor in this fascinating field in its very early period. Thus they both have personal experience in this domain in addition to their thorough and competent expertness. I am proud of the fact that the senior author, Dr. József Székely, was my former student and co-worker for many years working with me in several fields of psychophysiology and behavioral sciences. His outstanding erudition and creativity constitute a solid guarantee for the value of the present work. Similar qualification is valid for the junior author, Dr. András Rónai, a talented and devoted young pharmacologist. They both represent the most active and productive generation of Hungarian scientists, owing to whom Hungarian neuroscience became an important component of the mainstream of world brain research.

I highly recommend this book to all those physicians, psychologists, biologists, and other related specialists, who are driven by intellectual curiosity in order to get acquainted with the important and exciting field of neuromodulators of the brain.

Budapest, June 1982

Gyorgy Adam M.D., D.Sc.
Professor of Physiology
Member of the Hungarian Academy of Sciences

PREFACE

Since the late 1960s it is rapidly becoming clear that our classical conception on neuroregulation needs profound revision. It is no more tenable that the functioning of the excitable tissues (nerve and muscle cells) is regulated exclusively by locally, i.e., transsynaptically acting neurotransmitters, and only the activity of nonexcitable tissues is controlled by hormones (produced in remotely located endocrine glands and transported via the blood flow). Furthermore, it is already unequivocally proven that the dendrites and somata of nerve cells located on postsynaptic membranes or on the effectors are not the only sites of neuroregulation. As a matter of fact, an important locus of subtle regulation of signal transmission is located on the presynaptic portion of nerve cells which produce the neurotransmitters. A new group of endogenous neuroregulators has appeared on the horizon, which cannot be termed either classical neurotransmitters or hormones. These substances, mostly peptides, resembling in certain aspects the classical neurotransmitters, in other respects the peripheral hormones, are called nowadays neuromodulators. From year to year new ones are discovered or some previously isolated substances are proven to function as neuromodulators. With understanding the mode of their action our conception on chemical neuroregulation in the organism is getting more and more complex. Simultaneously, the development of relevant methods for studying receptor binding and biochemical events linked to receptors gave a closer insight into receptor-mediated processes.

The discovery of opiate receptors and their endogenous ligands (endorphins, enkephalins, etc.) represents a typical example of these recently burgeoning investigations. The authors of this book shared the enthusiasm of many fellow researchmen in this field since in a very early phase of these studies, they had the advantage of a fruitful cooperation with highly qualified preparative biochemist (headed by L. Gráf) and peptide synthetising (headed by S. Bajusz) groups. Moreover, they had already been working in the field of narcotic analgesics for some years before the most critical period of endorphin research (1976 to 1977). Consequently, this review cannot be regarded as a completely unbiassed description of certain events, as they represented personal experience for the authors themselves. Nevertheless, great care has been taken to remain as objective as possible.

The main purpose of this book is to give a detailed survey on the pharmacology of recently discovered endogenous opioid peptides and their synthetic analogues. The data compiled in these volumes will probably enable the academic scientists and those engaged in industrial research to become better acquainted with the results of endorphin research. It is hoped that the material will be useful also for those clinicians who are investigating the role of endogenous opioids in the pathogenesis of various diseases and the contingent therapeutic application of opioid peptides or their antagonists. Since the detection of formerly unknown neurohormones may always lead to the development of new drugs, special emphasis is given to the data on the structure-activity relationships. For those who are closely involved in the research of endogenous opioids, it is believed that the first-hand information is superior than any comprehensive treatise; furthermore it is hoped that they could offer some new aspects and also certain new data not accessible to them.

Reviewing the pharmacology of opioid peptides, primarily the biological aspects of endorphin research are discussed. Especially the data available on the in vitro and in vivo pharmacological properties of opioids and their clinical significance are reviewed in detail. As for the detailed description of biochemical isolation of natural opioids, the synthesis of their analogues, the physical and chemical properties of these compounds, the necessary references can be found in the corresponding chapters. In general, it was tried to quote all the relevant papers, but those published during the last one and a half years are not included due to the time-consuming nature of processing lengthy manuscripts like this one.

The first volume of the book is devoted to the description of in vitro and in vivo methods used in the opiate research. In this part the most characteristic pharmacological properties of classical morphine congeners are discussed in view of our present knowledge on the existence and multiplicity of opiate receptors. The pharmacological properties of opioid peptides are outlined in the second volume. The third volume contains the data on the structure-activity relationships and those on the clinical significance of endorphin research are discussed also there.

Thanks are due to several outstanding scientists for permission to reprint some very instructive illustrations from their original works.

The authors want to express their gratitude to Dr. Tibor Láng, Director of the Institute for Drug Research, Budapest, Hungary for his moral support on writing this book. Dr. Erzsebet Miglecz and Dr. Ilona Berzétei have rendered valuable help with collecting, arranging, and revising the material. Ms. Teréz Sági, Ms. Gizella Szegvári, and Ms. Tünde Tóth took part in the secretarial work.

Lastly the authors wish to thank their wives, Maria Székely and Dr. Ilona Berzétei for their tolerance, encouragement, and constant support to complete these volumes.

József I. Székely
András Z. Rónai

THE AUTHORS

József I. Székely, M.D., Ph.D., is Deputy Head of the Medical Department at the Institute for Drug Research in Budapest, Hungary.

Dr. Székely, born in 1937, received his medical training at the Semmelweis University Medical School in Budapest. After having obtained his first degree (M.D.) in 1961, he spent eight years at the Physiological Department of the same university (in the late 1960s as assistant professor). In the meantime Dr. Székely acquired the M.Sc. degree in psychology at the Eötvös Lóránd University in Budapest. He defended his thesis on the conditioning of evoked potentials in 1970. Dr. Székely joined the Institute for Drug Research in 1969.

He is co-author of about 60 papers more than half of them published abroad, i.e., in Western European and American journals. His recent research interests are related to the psychopharmacology, more closely to behavioural analysis of opioid peptides and other centrally acting pharmaca.

András Z. Rónai, M.D., born in 1947, received his first degree (M.D.) in 1971 at the Semmelweis University Medical School.

He started his pharmacological studies at the Department of Pharmacology of Semmelweis University Medical School in 1968 as an undergraduate working in Professor E. S. Vizi's team. After graduating he continued his activities at the same Department as junior lecturer. As a part of his postgraduate training, he spent a one-year fellowship at the Montefiore Hospital and Medical Center and Albert Einstein College (1974-1975) in the laboratory of Professor F. F. Foldes.

Late in 1975 he joined the Department of Pharmacology at the Institute for Drug Research. At present he is working as staff member in Professor Vizi's group at the Institute of Experimental Medicine of Hungarian Academy of Sciences.

Dr. Rónai, in addition to Hungarian scientific societies, is the member of European Neuroscience Association, International Brain Research Organization, and International Narcotic Research Club.

Dr. Rónai has published over 40 papers, with three exceptions in Western periodicals; besides, he is the co-editor of book entitled *Endorphins'78*. His major research interests are the pharmacology of narcotics, the pharmacological and physiological aspects of endogenous opioids, and the mechanisms of presynaptic regulation.

ABBREVIATIONS

β-LPH	Beta-lipotropin
α-EP	Alpha-endorphin
β-EP	Beta-endorphin
γ-Ep	Gamma-endorphin
E	Enkephalin
Met-E	Methionine-enkephalin
Leu-E	Leucine-enkephalin
DALA	
D-Met, Pro-EA	(2-D-Met,5-Pro)-enkephalinamide
FK 33-824	(2-D-Ala,4-MePhe,5-Met) 0 (-ol)-enkephalin

To our wives

Maria Székely and Dr. Ilona Berzétei

Opioid Peptides

Jozsef I. Szekely and Andras Z. Ronai

Volume I: Research Methods

In Vitro Analysis
The Most Characteristic In Vivo Effects of Opiates

Volume II: Pharmacology

The Mode Of Action of Opioids
Electrographic Analysis of Opioid Mechanisms
Analgesic and Behavioral Effects of Opioids
The Role of Endogenous Opioids in the Vegetative Regulation
Opioid Peptides and the Neuroendocrine Functions

Volume III: Opiate Receptors and their Ligands: Experimental and Clinical Aspects

Opiate Receptors
Detecion of Endogenous Ligands of Opiate Receptors
Regional Distribution of Enkephalins and Endorphins in the Brain and in the Peripheral Tissues
Biosynthesis, Release, and Biodegradation of Enkephalins and Endorphins
Structure-Activity Relationships
The Question of Metabolic Stability and Altered Binding Properties
The Pharmacological Profile of the Most Active Enkephalin Analogues
The Possible Role of Endogenous Opioids in Stimulation and Acupuncture-Induced Antinociception
Endogenous Opioids and Opiate Tolerance, Dependence, and Addiction
Clinical Investigation of the Opioid Peptides
Conclusions: Perspectives of Opioid Peptide Research

TABLE OF CONTENTS

Volume I

Chapter 1
In Vitro Analysis 1
A. Z. Ronai

Chapter 2
The Most Characteristic In Vivo Effects of Opiates 29
J. I. Szekely

Index 111

Chapter 1

IN VITRO ANALYSIS

A. Z. Rónai

TABLE OF CONTENTS

I. Introduction 2

II. Receptor Binding Assay (RBA) 2

III. Radioimmunoassay (RIA) 5
    A. The Specificity of RIA to Enkephalins 5
    B. The Specificity of RIA to β-EP 6
    C. RIA to β-LPH 6
    D. RIA to α-Endorphin 7

IV. Isolated Organ Technique 7
    A. Isolated Organs Sensitive to Opioid Compounds 8
        1. Preparations from Small and Large Intestine 8
        2. Preparations from the Urogenital Tract 9
        3. Preparations from the Cardiovascular System 9
    B. Other Isolated Organ Preparations from the Peripheral Autonomic and Motor System 9
        1. Isolated Nictitating Membrane 9
        2. Isolated Autonomic Ganglia 9
        3. Motor Nerve Terminals 9
    C. Isolated Organs Applied Most Frequently in Opiate Research 10
        1. Guinea Pig Ileum (GPI) 10
        2. Mouse Vas Deferens Preparation (MVD) 13
        3. Rat Vas Deferens Preparation (RVD) 14
        4. Cat Nictitating Membrane Preparation (CNM) 14
        5. Rabbit Ear Artery Preparation (ART) 16

V. Biochemical Methods 16

References 18

I. INTRODUCTION

The study of interaction of biologically active substances with specific binding entities in the organism had yielded in the developing of two important methods: the radioimmunoassay[1,2] and the receptor binding assay.[3,4]

Both methods have been introduced into[5-9] and are still widely applied in studying opioid mechanisms.

II. RECEPTOR BINDING ASSAY (RBA)

Basically, the receptor binding assay covers the detection of the binding of a properly chosen labeled substance (primary "ligand") to its specific "receptor(s)" under properly chosen experimental conditions.

The theoretical and methodological implications of ligand-receptor interaction have been treated by several review and comprehensive studies.[4,10-16] Some general considerations and also the specific issues related to opiate receptors will be discussed below.

The "receptor", as traditionally conceived, possesses (1) the capacity to bind its "ligand(s)" with high affinity and specificity as well as (2) the capacity to activate cellular processes. The two functions are distinct and often dissociable; the method to be discussed below is concerned primarily in characterizing the former (i.e., "recognition") function.

When dealing with binding entities present in a biological system for a given substance ("ligand"), one must decide upon the criteria to identify the binding site(s) as biologically/pharmacologically relevant receptor(s). Thus, the binding of the "ligand" to the putative "recognition" macromolecule to be identified as "receptor", must exhibit (1) high affinity, (2) saturability, (3) reversibility (there are exceptions, though), (4) selectivity, i.e., the binding sites must distinguish the biologically/pharmacologically relevant ligands, (5) the tissue distribution of these binding sites has to be consistent with the biologically acknowledged targets for the given family of active substances, (6) the binding isotherm for a given substance has to be related to the concentration range over which it is known to exert its biological/pharmacological action, and (7) the relative affinity of analogs of the biologically/pharmacologically active substance has to be in keeping with their biological/pharmacological potencies. As it is apparent, most criteria (4 to 7) presume well-defined knowledge on the biological/pharmacological properties of the substances under study; some discrepancies, however, might find their explanation in the scarcity of our knowledge (in relation of opioid mechanisms, see Reference 17). In general, however, only when such criteria are met (and, if exceptions are present, reasonable explanation can be given), is justified to assume that the binding entity we are dealing with, does indeed represent "the" receptor.

As a matter of fact, gathering of information by pharmacological tools must precede even the postulation that specific receptor for a family of substances does exist in the organism. The history of opiate research provided copious evidence to such a postulate,[18] the accumulated pharmacological data rendered even possible to construct a well-defined and relevant receptor model as early as 1954[19] (see also Chapter 1, Volume III).

The production of highly potent opiate agonists and antagonists like etorphine, naloxone, diprenorphine, etc., provided the opportunity to select suitable candidates for labeling to high specific activity, to obtain "tracers" with high affinity to the postulated tissue receptors.[5-8]

The knowledge that the steric configuration of opiates is important for their pharmacological effect[19] and also the possession of several enantiomeric pairs of morphine surrogates which differed in their pharmacological potency by orders of magnitude, offered tools to test the stereospecificity of binding, since, as it has been forwarded by Goldstein et al.,[5] the receptor binding should be stereospecific. The saturability and stereoselectivity of opiate binding to brain membranes has been shown by demonstrating that the "tracer" can be displaced from the binding sites by excess of pharmacologically active "cold" enantiomer but not by the inactive one.[6-8] Stereospecific binding of opiates occurs also to pharmacologically irrelevant objects, e.g., glass fiber filters;[20] it has been proved that the saturable, stereospecific opiate binding that had been successfully demonstrated in brain membrane preparations does represent the interaction with pharmacologically relevant receptor structures.[20,21]

The technical details of securing that the specific binding process be detected have been treated in the early papers and have recently been reviewed by Höllt.[15]

The proper choice of labeled ligand ("tracer", "primary ligand"), the biological system where the binding is measured, the technique to separate bound and free ligand, the composition and temperature of incubation media, and the time constant of assay are designed to ensure that relevant and reproducible data be obtained.[15]

The experiments which are based on the detection of binding of labeled ligand to its specific tissue receptor(s), may serve two purposes: (1) to characterize the ligand-receptor interaction itself (RBA) and (2) to visualize the localization of specific receptor(s) in the tissues, to clarify their relation to anatomical structures and thereby to characterize their distribution in the organism. The latter field of application in relation of opioid receptors is discussed in Chapter 1, Volume III; it should be mentioned that the RBA could also be designed so as to earn information on the distribution of specific binding sites.

The classical way to characterize the binding of a ligand to its receptor in the RBA is to perform saturation or displacement experiments. Thus, either increasing concentrations of a labeled form of the ligand or a fixed concentration of radioactive and various concentrations of cold ligand are incubated with receptor-containing preparation and the specific binding of tracer is measured. By the aid of RBA, the following questions could be answered:

1. The affinity of a ligand to the receptor(s)
2. The number of specific receptors in a sample
3. The properties of recognition macromolecules

The latter point covers the problem of homogeneity and/or heterogeneity of receptors in a given experimental system; the possible cooperativity between the specific binding entities and the possible presence and character of allosteric site(s) where the receptor-ligand interaction can be modulated.

To answer these questions, one can relate the concentration of specifically bound ligand (B) to the concentration of free ligand (F). The binding data can be represented graphically in several ways; the plots with manipulated primary data are intended to linearize the graphs for special cases of receptor — ligand interaction. Thus, from the linear plots some parameters characterizing the binding (e.g., the dissociation constant, K_D and the total number of specific binding sites, R_T) could easily be determined graphically; also the deviations from the "special" premises, could conceivably be detected.

Some of the possible graphical representation forms are listed below, after Boeynaems and Dumont:[14]

1. Direct plot, adsorption isotherm, plotting B, ordinate vs. F, abscissa
2. Semilogarithmic plot (B vs. lgF)
3. Double-reciprocal plot, Lineweaver-Burk[22] (1/B vs. 1/F)

4. Half-reciprocal plot, Scott[14] (F/B vs. F)
5. Scatchard plot[23] (B/F vs. B)
6. Hill plot[24]

$$\lg \frac{B}{(R)_T - B}$$

7. Proportion graph, Baulieu and Raynaud[25]

$$\lg \frac{B}{B + F}$$

8. Double logarithmic plot, Thompson and Klotz[26] (lgB vs. lgF)
9. Boeynaems and Swillens[14] (B²/F vs. B)

The Scatchard[23] and Lineweaver-Burk[22] plots were designated to be linear in the case of interaction between a ligand and one homogeneous set of noninteracting binding sites. The Scatchard graph is the most widely used one in the field of binding studies. Provided that the experimental conditions (in relation of opiate binding see Reference 27) and the evaluation[28] are handled properly, it is a sensitive though often ambiguous evaluation method.[14] For detecting and characterizing cooperativity the Hill-plot[24] is a rather frequented though also not an unambiguous approach.[14]

The graphs can be interpreted validly only if the data afforded by the saturation experiments are not distorted by methodological errors: thus, B must be the true measure of bound ligand concentration, and F the true measure of free ligand concentration. Furthermore, the measurements have to be performed in a true state of equilibrium. These methodological pitfalls have to be kept in mind even more attentively if the ligands we are working with might be susceptible to degrading enzymes present in the assay system, as it is the case of numerous opioid peptides.

In the displacement experiments, the most frequently calculated data to characterize quantitatively the interaction of cold ligand with the receptor are the IC_{50} and the inhibition constant (K_i) values.[29] The latter can be calculated from the equation

$$K_i = \frac{IC_{50}}{1 + [L]/K_D}$$

where [L] is the concentration of labelled primary ligand ("tracer") and K_D its equilibrium dissociation constant.[30]

In relation of opiate receptors, the existence of distinct, heterogeneous receptor types has been proposed both from the pharmacological[31-34] and binding experiments.[32,35-40] Though several types of labeled opiate ligands are at hand, both from the morphine surrogates (morphine, dihydromorphine, naloxone, naltrexone, etorphine, diprenorphine, ethylketazocine, SKF 10047) and from the opioid peptide family (Met-E, Leu-E, β-EP, DADL, FK 33-824)[6-8,29,32,35-43] of the numerous types of opioid receptors proposed by the pharmacologists, i.e., μ,[31,32] κ,[31,32] σ,[31] δ[29,32] ε,[33] and dynorphin,[34] only two subclasses could be identified unambiguously in the receptor binding assay, which may correspond most closely to the so called μ-[29,32,40] and δ[29,32,40] receptors (see also Chapter 1, Volume III).

The phenomenon of cooperativity has also been demonstrated in opiate receptor binding[44] as well as the allosteric regulation of opiate binding by cations[37,44,45] and by guanyl nucleotides[37,38] (see also Chapter 1, Volume II and Chapter 1, Volume III).

III. RADIOIMMUNOASSAY (RIA)

It has early been realized[46,47] that the sensitivity and specificity of antigen-antibody reaction offers an outstanding tool for detecting minute amounts of substances present in the body fluids and various tissues. The sensitivity of detection of biologically active materials by the aid of radioimmunoassay, a method, based on the above-mentioned principle, surpasses even that of the most sensitive bioassays. Its specificity — in a sense — is also superior to the measurements based on biological determinations. At the moment of discovery of endogenous opioid peptides, RIA was routinely applied for the quantitative determination of various substances in the organism, like peptide — and steroid hormones, vitamins, drugs and metabolites thereof, toxins, etc. Soon after the discovery of the primary structure of Met-E[48] and other opioid peptides structurally related to β-LPH[49,50] — the knowledge of which is a prerequisite for RIA — the method has been introduced also in the research of endorphins.[51,52]

To prepare immunogens for raising antibodies usually in rabbits or guinea pigs, the opioid peptides are conjugated with bovine serum albumin,[51,53] thyroglobulin[52,54,55] or ovalbumin.[56] To achieve high antibody titers, adjuvants and "booster" injections are applied. The radioiodinated derivative of parent opioid peptide is prepared to serve as "radiotracer"; the Tyr-moiety present in these substances provides the feasibility of labeling. There are slight methodological variations in relation of separating the free peptide in the RIA;[52,54,55] both the double antibody precipitation[52] and the charcoal adsorption[54] methods gave comparable results.

Thus, radioimmunoassays have been developed to enkephalins[51,55-62] β-LPH,[63-69] β-EP[52-54,57,58,70,71] and α-EP.[52,70] The lower limit of detection, depending on the peptide was reported to be in the 1 to 100 pg/test tube range; there are considerable variations in the values reported by different groups even for the very same peptide. For each "tracer" the specificity of the assay must be meticulously tested. The specificity is characterized by the lack (or low degree) of cross-reaction with structural congeners of the tracer as well as with other known substances, for which the possibility of cross-reacting may emerge. This indirect way of specifying represents one of the pitfalls of the method as compared to chemical of biological determinations; the possibility of interference by unknown substance(s) in the assay system has to be kept in mind upon interpreting the results. In the bioassay, where not a structure but an effect is detected, the specificity of determination can be checked by using specific antagonist(s) (provided, that such substance(s) is (are) available; thus, in such cases the specificity can safely be characterized in relation of biological effect. On the other hand, the structural determinants of antigenic properties and of biological effectiveness do not necessarily coincide in a given substance (see below); this fact is another possible source of misinterpretations.

A. The Specificity of RIA to Enkephalins

The RIA generally used for the detection of enkephalins, differentiates well by the C-terminus of pentapeptides. The cross-reactivity between Met- and Leu-E was reported to be by the different groups of authors[51,55,56] below 5 to 7%, in both directions. Longer β-LPH-related sequences,[55,56] enkephalin fragments[51,55,56] and a number of substances structurally unrelated to opioid peptides (e.g., norepinephrine, GABA, acetylcholine, cyclic AMP, etc.)[51,55,56] or nonpeptide opiates (morphine, methadone, naloxone, naltrexone, etc.)[51,55,56,70] show even lower degree of cross-reactivity or no cross-reactivity at all.

To promote the detection of endogenous, C-terminally extended enkephalin congeners (Met-E-Arg⁶, Met-E-Arg⁶, Arg⁷, Met-E-Arg⁶, Phe⁷, Met-E-Lys⁶). Rossier et al.[72-74] have developed RIA for "reading" the N-terminus of these substances; the selectivity of this assay, of course, was less than those of the above mentioned RIAs.

B. The Specificity of RIA to β-EP

The antigenic determinants of β-EP (1-31) reside in the C-terminal part of the peptide.[52,54,70,71,75-77] Thus, the immunoreactivity of C-terminally shortened congeners of β-EP falls drastically only when the peptide chain is shorter than β-EP (1-23).[52,54,75,77] The sequence portion β-EP (22-31) is responsible for being specifically recognized by antisera raised against β-EP. Since β-EP-(18-31) has been found to be inactive in the RIA to β-EP[30] it appears to be justified to assume that the β-EP antiserum recognizes a specific conformation of β-EP rather than a particular sequence portion of it.[52,70,75]

Several consequences emerge from the fact that the specificity of reaction is provided by the C-terminal part of β-EP. Thus, the antisera hitherto developed to β-EP react to a considerable extent also with Leu⁵- β-EP⁷, β-LPH[52,54,65,70] and with the 31K common precursor of ACTH and β-EP (pro-ACTH) endorphin[52,71,78] (see also Chapter 4, Volume III).

Since it is very important to decide that in a given biological system β-EP or β-LPH is detected by the aid of RIA, Krieger et al.[65] prepared antibodies directed to the NH₂-terminal portion of β-LPH (see below), giving no cross-reactivity with β-EP.

A further consequence is that the antibody raised against β-EP reacts almost fully with β-EP(2-31)[52,70] and does react also with β-EP (6-31) and (10-31);[70] these fragments, however, are devoid of any significant opioid activity. Also, N-acetyl (Tyr¹)-β-EP shows full reactivity with β-EP antisera,[79-81] while it is inactive as opioid agonist to β-EP.[79-82]

In certain brain areas this form has been found to be the dominant and not genuine β-EP, as it had been thought previously.[80] Since N-acetylation does not occur extracellularly,[79] moreover, the N-acetylation step is likely to be closely linked with the post-translation formation of β-EP[81], it means that at these brain sites the tissue stores are made up mainly of an endogenous opioid structure practically devoid of opiate activity (see also Chapter 4, Volume III). It is easy to realize that the theoretical implications of this fact may modify some theories based on the presence of biologically fully active β-EP in these regions.

A further disparity between the structural determinants of biological activity and immunoreactivity was found in single-amino acid-deleted analogs of camel β-EP: depending on the biological assay, (des-Leu¹⁴)-β_c-EP and (des-Asn²⁰)-β_c-EP retained enhanced, full, or just slightly reduced biological activity as compared to β_c-EP, while the cross-reactivity with antibodies raised against β-EP, was lost.[83]

The "favorable" consequence of C-terminal determinants of immunoreactivity is that the antibody to β-EP does not cross-react appreciably with α- and γ-endorphins and Met-E.[52,54,70]

The lack of cross-reactivity was found for human ACTH and its fragments (1-17 and 1-23); in great excess, porcine ACTH (1-39) showed some cross-reactivity, probably due to contamination (the antibody was directed to β_h-EP).[54]

C. RIA to β-LPH

The antiserum developed by Krieger et al.[64,65] to β-LPH contained two distinct binding sites, one with an affinity for β- and γ-LPH and one with an affinity for β-LPH and β-EP. If the COOH-terminal-reactive antibodies were exhausted with β-EP,[65] then the recovered antiserum exhibited selectivity for the NH₂-terminal portion of β-LPH with no further cross-reactivity with β-EP.

D. RIA to α-Endorphin

The antiserum raised to α-endorphin ("β-EP-(1-16)")[52,70] recognizes the COOH-terminal portion (10-16) of the peptide. It does not react with Met/Leu-E, morphine, heroin, methadone, naloxone, naltrexone.[52,70] The cross-reactivity with γ-EP, β-EP (and Leu[5]-β-EP) and β-LPH is rather low but clearly detectable;[52,70] surprisingly high is the cross-reaction with Met(O)^5-β$_h$-EP.[70]

IV. ISOLATED ORGAN TECHNIQUE

The history of opioids is closely connected with examination of their neural actions. The therapeutically exploitable biological properties and also the unwanted side effects of opiates originate mostly from their action on the central nervous system.[84] However, as it has early been realized, they also affect certain autonomic functions by acting upon the peripheral elements. The latter experience led to the development of isolated organ preparations which are specifically sensitive to narcotics.

The use of isolated organs in studying the effects of opiates can be traced back as far as 1917, when Trendelenburg[85] demonstrated that the peristaltic reflex elicited by distention of the lumen of the isolated ileum of guinea pig was inhibited by low concentrations of morphine. The isolated organ technique using mostly smooth-muscle preparations containing neuroeffector junctions sensitive to opiates helped to establish more closely the structure-activity relationships of natural opiates, semisynthetic and synthetic narcotic derivatives as well,[86-89] than did the former analgesic measurements.[90-92]

In the critical period of the detection of endogenous opioid peptides, i.e., between 1971 to 1975, the research strategy built on the application of isolated organ techniques[93-95] contributed to a considerable extent to the discovery of enkephalins.

This technique has several advantages as compared to the in vivo methods used for studying the opioid effects. The most frequently quoted one is that the disturbing influences of absorption, distribution, biotransformation, and excretion of the drugs are almost completely excluded. Moreover, — far from stating that the isolated organs are simple assay systems — due to the "isolated" nature of testing systems many sources of error can be avoided (e. g., compensatory mechanisms secondary to the drug effects may not complicate the interpretation of the results). The transmission at autonomic junctions on which the opioids exert their effect can be correlated more or less directly to a well-defined function (i.e., contraction or relaxation of smooth muscle, due to the alteration of local neuronal activity); and these effects can reproducibly be measured and quantitated. The local neuronal network — both the intrinsic elements and the neuronal input — can be better identified[96,97] and their functional correlation could be determined easier than in most cases when working with CNS tissues. A further advantage of the isolated organ technique is that the neuronal pools involved are more or less homogeneous. The application of isolated organ techniques in opiate research covers several domains. The first and of relatively lesser importance is when measuring an opiate effect in an organ one draws conclusion on the possible significance of the effect in the function of the organ studied. Far more important is to use the isolated organs as models of CNS functions and to correlate the results of the actual measurements with the data of in vivo assays.[86,89,98] Thus, one can follow structure-activity relationships in a series of compounds, and predict the potential in vivo efficacy for a special CNS function. One can also construct models for some characteristic CNS effects of opiates like tolerance and dependence;[99] these experiments would offer a closer insight into the mechanism of these phenomena. Since the neural elements of certain isolated organs contain opioid peptides which can be eliberated by different types of stimuli,[100,101] one can also study the mechanism of release and biosynthesis of endogenous opioids. Furthermore, using various ligands this technique — besides the receptor binding assay and receptor isolation — characterization — helps to elucidate the heterogeneity of opiate receptor mechanisms and to differentiate between the classes (types) of opioid receptors.

The role of isolated organs in opiate research has been reviewed both before[89] and after the discovery of enkephalins.[99] To avoid unnecessary overlaps with the previous reviews the present chapter will follow the next composition: first, in a general overview, the preparations sensitive to opiates and/or opioid peptides will be listed. The properties of the five most frequently used preparations shall be discussed in detail, such as the guinea pig ileum, mouse and rat vas deferens, cat nictitating membrane, and rabbit ear artery. Special emphasis will be laid on the demonstration of similarities and differences of opioid receptors in these isolated organs.

A. Isolated Organs Sensitive to Opioid Compounds
1. Preparations from Small and Large Intestine

Morphine has a smooth muscle contracting effect in isolated canine small intestine.[102] This stimulatory action is indirect, i.e., it is due to an effect on nerve elements affecting smooth muscle activity. A variety of opioid peptides also contract segments taken from different parts of rat intestine;[103] the effects are most prominent in colon and rectum. The specificity of action was also proved for the latter organ by kinetic analysis. In preparations taken from guinea pigs these agents relax rather thank stimulate the small intestinal portions; different parts of large intestine are, however, contracted by opioid peptides. The stimulatory effect is much weaker as compared to the parent segment taken from rat, and in most cases it is of transitory character.

The peristaltic reflex elicited by the distension of the lumen of isolated ileum is inhibited by morphine and a variety of narcotic analgesics in guinea pigs,[85,104,105] but not or just moderately in rabbits.[106]

Contractions of the longitudinal muscle of guinea pig ilea evoked by different agents like serotonin,[105,106-108] nicotine,[105] "Darmstoff",[109] angiotensin,[110] cholecystokinin, pentagastrin,[111] arachidonic acid peroxide[112] and $BaCl_2$[105,106] can also be inhibited by opiates.

Contractions to exogenous acetylcholine and histamine, however, are not influenced by morphine and its congeners.[98,113] The circular muscle of the very same organ is much less responsive to contracting agents in experimental conditions when it is constricted by 5-HT, nicotine or histamine, the stimulatory action of these drugs can be counteracted by morphine.[113]

The electrically stimulated intestinal preparations are the most frequently used ones in the investigation of opiate effects. The mode of stimulation and the parameters are chosen such that it affects mostly — if not exclusively — the nerve elements. The cholinergic transmission from the myenteric plexus to the longitudinal muscle is depressed by morphine in the guinea pig ileum,[98] but not in the rabbit or rat.[114] The guinea pig ileum preparation is very suitable for studying the kinetics of opioid action.[89,115] Besides, it is also sensitive to opioid peptides of different kind.[95,116-118] The preparation has high opioid peptide content;[119] furthermore, these endogenous opioids can be eliberated from the organ (see also below) by nerve stimulation using properly chosen parameters.[100,101]

On the other hand, morphine has no influence on the adrenergic inhibitory mechanisms either in guinea pig ileum[120] or taenia coli[121] preparations. Morphine, however, depresses some nonadrenergic inhibitory responses in guinea pig taenia coli.[121] Morphine also exerts inhibitory effect in rat vagus — stomach[122] and vagus — esophagus preparations.[123]

2. Preparations from the Urogenital Tract

The electrically stimulated vas deferens preparations constitute the other most frequently applied group in investigating opioid effects. Morphine and its congeners inhibit the contractions of the longitudinal muscle elicited by nerve stimulation in the mouse vas deferens,[88,89,124] but not in the rabbit, guinea pig, rat, cat, hamster, or gerbil.[88]

On the other hand, the rat vas deferens, while being insensitive to morphine and its synthetic surrogates, displays specific and rather selective sensitivity towards endorphin and some synthetic opioid peptides.[125,126] This property of rat vas deferens is shared (at least partly) by similar preparations from rabbits but not from cats or guinea pigs.[125] Finally, it is worth mentioning that the vas deferens of young (10 to 16 days old) rats are sensitive also to morphine-like compounds;[127] the loss of responsiveness takes place during postnatal maturation.

In rat uterine preparations morphine antagonizes the twitches elicited by PGE_2 administration.[128]

As for the effect of opiates on some parts of the gut and urogenital tract *in situ* we refer to the competent review of Weinstock.[129] In general, morphine increases the tone and the contraction of muscles both in the ureter and in the bladder. This effect is most pronounced on the sphincters. The smooth muscle contracting action of morphine could also be detected in excised ureters of guinea pigs and rabbits.[130,131]

3. Preparations from the Cardiovascular System

The transmission from the vagus nerve to the sinoatrial node is morphine sensitive in the rat and rabbit but not in the guinea pig.[132] Except in very high concentrations morphine does not affect the release of norepinephrine from the rabbit isolated heart.[133] The noradrenergic transmission in rabbit ear artery preparation is insensitive to morphine but it is highly susceptible to the inhibitory action of Met-E[51] and certain enkephalin analogs.[135,136] The transmission in the portal vein of the same species, however, is not influenced by morphine.[120]

In the light of findings on rat vas deferens,[125,126] the specific sensitivity of certain cardiovascular preparations towards opioids of peptide nature, ought to be tested systematically.

B. Other Isolated Organ Preparations from the Peripheral Autonomic and Motor System

1. Isolated Nictitating Membrane

Since the early experiments of Trendelenburg[137] carried out *in situ* in cats, it is known that the preparation should be classified among the organs containing morphine-sensitive neuroeffector junctions. The isolated, electrically stimulated medial muscle of the cat nictitating membrane[138] is also susceptible to the specific inhibitory action of opiates.[120,135,136,138-143]

Both the in vivo[144] and in vitro preparations[135,136,142,143] are suitable for studying the effects of opioid peptides.

2. Isolated Autonomic Ganglia

Isolated frog sympathetic ganglia have been reported to be sensitive to opiates, using electrophysiological tools for the demonstration of action[145-147] (see Chapter 2, Volume II).

3. Motor Nerve Terminals

Opioid substances[148] were found to act also at the motor nerve terminals in mammalian striated muscle, though this class of action is much less extensively documented than the one at the peripheral autonomic junctions. The opioid peptides Met- and Leu-E reduce the quantal release of acetylcholine in mouse phrenic nerve-diaphragm preparations without acting at the postsynaptic elements.[148]

C. Isolated Organs Applied Most Frequently in Opiate Research

All the preparations to be described below are basically nerve-muscle preparations, innervated by the peripheral autonomic nervous system. In these isolated systems opiates inhibit the contractions of musculature elicited by indirect (nerve) stimulation through the reduction of the amount of transmitter released. Whether they act at the nerve terminals or at some nerve elements proximal to it is not decided yet (see Chapters 1 and 2, Volume II).

The specificity of the inhibitory action can be demonstrated by using specific, competitive opiate antagonists like naloxone, naltrexone, levallorphan, etc.[87-89,92,149] The agonist potencies of opioid compounds can be characterized by the 50% inhibitory doses (ID_{50} or IC_{50}). To quantify the action of a competitive antagonist the pA_2 or K_e value[87,150,151] can be determined. By definition, the K_e (equilibrium dissociation constant) value indicates the molar concentration of antagonist shifting (virtually) the agonist dose-response curve to the right in parallel fashion by the factor of 2.0, while pA_2 is its negative logarithm. The pA_2 value can be derived graphically from the Schild-plot;[150] both parameters can be calculated from the equation[87,151]

$$K_e = \frac{a}{DR - 1}$$

where "a" is the molar concentration of antagonist and DR (dose ratio) is the measure of the virtual shift of the agonist dose-response curve to the right, in the presence of a given concentration ("a") of antagonist. The DR values can be obtained by dividing the agonist ID_{50} (IC_{50}) values measured in the presence by the one determined in the absence of antagonist.

If, as it is the case for numerous morphine congeners,[152] an opiate compound possesses both agonist and antagonist properties, then the ID_{50} (IC_{50}) and K_e values can be obtained according to the method elaborated by Kosterlitz and Watt.[87]

These parameters in addition to characterizing the potencies of the agents in question can be used for characterizing also the receptors where these agents exert their action. If the order of agonist potencies of a number of opiate agonists is markedly different in different opiate-sensitive assay systems, then these systems are assumed to contain receptor populations of different character.[117,153,154] If a given antagonist antagonizes the actions of various agonists (or various actions of the same agonist) with significantly different potency — under experimental conditions where the comparison is justified — then the receptors where these agonist effects are exerted, are thought to be heterogenous.[117,153,154]

1. Guinea Pig Ileum (GPI)

Small intestinal muscle cells in general, receive a multiple innervation by cholinergic, adrenergic and "purinergic" nerves.[155] Enkephalin-, Substance P-, VIP-, and 5-HT-, containing nerves have also been demonstrated.[97,156,157]

It is generally accepted that the parasympathetic cholinergic outflow to the intestine has an excitatory motor action on both longitudinal and circular muscle layers (except in the region of sphincters where it is, as a rule, inhibitory). In the guinea pig ileum, no cholinergic fibers enter the longitudinal muscle.[158] Cholinergic terminals, containing characteristic vesicles lie at the border of the myenteric plexus opposite the muscle. Separation between autonomic nerves and smooth muscle of less than 100 nm are rare in the longitudinal muscle coat of intestine,[159] whereas close (20 nm) junctions are common in the circular muscle coat.[160]

The noradrenergic and "purinergic" innervation is of inhibitory function;[161-164] this action exerted in minor part directly on the smooth muscle,[165,166] but mostly by modulating the release of excitatory transmitter acetylcholine.[161,162] Endogenous opioids in the small intestine mediate inhibitory processes.[100,167,168] VIP- and SP-containing neurons were shown to connect the two ganglionated plexuses in guinea pig intestine;[97,156] they are assumed to mediate local functional circuits.[96,97,156] PGE$_2$ had also been shown to be involved in the modulation of acetylcholine release in isolated guinea pig ileum.[169]

Basically three types of measurements can be performed in GPI preparation to investigate opiate effects. One can record either electrophysiological phenomena (Chapter 2, Volume II) or measure the release of transmitters, or detect mechanical responses of the organ elicited by different kinds of stimuli.

The first two methods are applied mostly to determine the mechanism of action of opioids, while the latter is the most suitable one for studying structure-activity relationships which by no means would mean that this kind of measurement cannot be applied for theoretical analysis of opiate effect.

The investigator can detect the coordinated responses of circular and longitudinal musculature (i.e., the peristaltic response)[111,167,168] or the contraction of the longitudinal muscle either in whole ileum segments (with or without the mesenteric attachment) or in a strip — preparation containing only the longitudinal muscle and the Auerbach plexus.[86,87,89,98,116-118]

The most frequently used stimulation mode to elicit the mechanical responses of longitudinal musculature is the electrical stimulation either in the form of transmural stimuli or as field stimulation.[86,87,89,98,116-118,161] The parameters of stimuli are chosen such to affect the nerve elements and not directly the smooth muscle. Thus, the contractions of muscle are evoked indirectly, through the eliberation of excitatory transmitter acetylcholine. Morphine-like substances depress the release of acetylcholine,[98,170,171] which is reflected by the proportional diminution of muscle twitches. The inhibition is apparent at low-frequency stimulation or even at higher frequencies provided that the number of impulses in a train of stimuli is not too high.

The properties which make the organ particularly suitable for structure-activity studies are as follows:

1. The relative inhibitory potencies of numerous opiates (alkaloids, semisynthetic, and synthetic narcotics) in this system correlate well with their relative potency for analgesic action in man.[89,172]
2. The relative inhibitory potencies of agonists or the relative antagonist capacities of antagonists correlate closely with their order of affinity demonstrated in the receptor binding assay performed in homogenates of the same preparation, using ^{3}H-naloxone as primary ligand.[173,174]
3. Both the binding capacities and the relative agonist/antagonist potencies of opiates in guinea pig ileum can be corresponded well with their relative affinities to the opiate receptors in brain homogenates, judged from the displacement of ^{3}H-naloxone in the receptor binding assay.[89,173,174]

From the very beginning of the research of opioid peptides, the preparation has routinely been applied to detect the opiate effects of this new type of opioid substances. As it became apparent, however, the guinea pig ileum preparation alone is much less useful to reveal the characteristics of the action of opioid peptides than in the case of classical morphine congeners. It has been established[116,117] that a group of opioid peptides is capable of interacting with a class of opioid receptors, termed as "δ" which cannot be characterized functionally in guinea pig ileum (see also below and in Chapter 1, Volume III). It means that certain natural opioid peptides like Met- and Leu-E, which are capable of exerting approximately 25 to 60 times stronger specific opiate agonist effect than normorphine (morphine) in certain isolated organs, e.g., mouse vas deferens[116-118] (see below) are equipotent with or less potent than normorphine (morphine) in guinea pig ileum. Furthermore, while in mouse vas deferens approximately ten times higher dose of naloxone/naltrexone is required to antagonize the effects of enkephalins than is necessary against normorphine,[117,175,176] in guinea pig ileum these opiate antagonists counteract the actions of pentapeptides and normorphine with nearly equal potency. Inspecting, however, more closely the K_e values of the above-mentioned antagonists determined against peptide- and nonpeptide opioid agonists in guinea pig ileum, minor differences can be detected also in this isolated organ.[175,176]

The "dominant" opioid receptor type in the guinea pig ileum longitudinal muscle-myenteric plexus preparation, based on isolated organ technique, has been claimed to be "μ"[117] (see also Chapter 1, Volume III). In addition, the so called "χ"[153,154] receptors could also be detected by functional measurements.[117] As it was mentioned above, no significant amount of the so called "δ" receptors[117] could be suggested to be present in this preparation, based on the results of isolated organ experiments. In the receptor binding assay, however, carried out in the homogenate of longitudinal muscle-myenteric plexus[177] a considerable proportion of opioid receptors could be identified as "δ". To interpret this discrepancy further experimental analysis should be performed. Based on the very preliminary data obtained with a novel endogenous opioid peptide, dynorphin (1-13),[178] the presence of a further receptor type (defined as such lacking a better explanation) might be guessed in this isolated organ.

Opiate tolerance,[179-184] "acute" tolerance[98,185] and opiate dependence[186-189] can also be modeled in isolated guinea pig ileum preparation (for review see Reference 99). Opiate tolerance in isolated ilea of animals treated chronically with tolerance-inducing opioids is characterized at the first sight by a strongly reduced capacity of opioid agonists to depress the electrically induced, nerve-mediated muscle contractions.[99,179-184] There was a decrease in the apparent affinities of opiate antagonists in morphine tolerant preparations, determined by isolated organ technique.[182,183] Tolerance to opiate agonist effect was not accompanied, however, by a change in the affinity or the number of stereospecific opiate-binding sites in homogenates of myenteric plexus-longitudinal muscle preparations, as compared to the nontreated controls[183] in the receptor binding assay.

Tolerance to morphine was reported to be accompanied by a supersensitivity which exert their excitatory action mainly indirectly, i.e., through nerve elements (e.g., 5-HT, nicotine, KCl), while the contractile effect of acetylcholine, according to the majority of authors[99,179,184] remained unaffected.

Acute tolerance to morphine could also be developed in guinea pig ilea, though it cannot be so invariably generated then by chronic opiate treatment.[99,185] As it has been reported by Shoham-Moshonov and Weinstock,[185] the feasibility to develop acute tolerance in the preparation showed a distinct seasonal variation.

The dependent state in opiate-treated guinea pig ileum is characterized by a tetrodotoxin-sensitive (i.e., nerve-mediated) smooth muscle contracture upon challenge with naltrexone.[99,186-189] The naloxone-elicited "withdrawal" contracture is mediated by at least two neuronal systems: one involves acetylcholine and subsequent activation of muscarinic cholinergic receptors,[186] while in the generation of other, atropine-resistant component, serotonin[188] and substance P[189] appear to be implicated.

2. Mouse Vas Deferens Preparation (MVD)

The mouse vas deferens receives a dense and intimate adrenergic innervation[190] from close-contact varicosities on nerve processes.[191] Similarly to the vas deferens of other species, e.g., guinea pig,[192] there is a dispute whether or not the motor innervation is of noradrenergic nature. The experimental data seem to favor the assumption that the noradrenergic nerve supply is responsible for the motor innervation of the organ.[193,194] Ganglion cells are extremely sparse in the wall of mouse vas deferens;[194] pharmacological evidences also suggest[194] that the generally applied procedures for nerve stimulation affect postganglionic elements. The release of norepinephrine is controlled by presynaptic α-(α_2)-adrenergic receptors.[195] It is not fully decided whether dopamine receptors are[135,136] or are not[196] involved in the regulation of neuromuscular transmission.

The morphine-sensitivity of this preparation was established by Henderson et al.[124] Opiates and opioid peptides depress the neuromuscular transmission by reducing the release of norepinephrine,[124,194,197] acting at specific opiate receptors on the nerve terminals.[88,124,194,197]

The mouse vas deferens preparation can be applied for studying structure-activity relationships of morphine congeners;[88,89] it proved to be even more suitable for the examination of effects of opoid peptides.[93-95,116-118,175]

Morphine and its congeners in general, are less potent agonists in mouse vas deferens than in guinea pig ileum[88,89,116,117] though our group found normorphine nearly equipotent in the two isolated organs.[118,135,136] The discrepancies might be explained by the strain differences in the sensitivities of vas deferens to opioids;[198-201] these differences may involve variations in the relative proportion of the so-called "μ" receptors in this preparations. The mouse vas deferens displays prominent sensitivity to the agonist action of endogenous opioid peptides like Met/Leu-E.[95,116-118,201-203] These natural pentapeptides are approximately 10 to 60 times more potent agonists in this preparation than in guinea pig ileum.[95,116-118,201-203] β-EP, however, was found to be nearly equipotent in these two isolated organs.[116-118,202,203] It is worth to note that the current strength of the electrical stimulation applied has been shown to influence differentially the sensitivity of vas deferens to different opioid peptides.[204]

The differential behavior of opioid peptides in mouse vas deferens is apparent not only from their relative agonist potencies as compared to guinea pig ileum but also from their antagonizabilities by opiate antagonists. As it was mentioned above, approximately 10 times more naloxone/naltrexone is needed to antagonize the agonist actions of Met/Leu-E and β-EP in this preparation[117,175,176] than is necessary against normorphine. This property of the above-mentioned natural opioid peptides is shared by certain synthetic derivatives, while other synthetic analogs became normorphine-like in this respect.[136]

These peculiarities of opiate mechanisms in mouse vas deferens have prompted Lord et al.[117] to propose the existence of a novel type of opioid receptor in this organ, designated as "δ" by the authors. The existence of the so-called "δ" binding sites has also been demonstrated by the aid of receptor binding assay in homogenates of the preparation.[177] This is the "dominant" receptor type in this organ;[117] there are fewer "μ"[117,177] and very few, if any "$\varkappa$" receptors.[117] The existence of specific receptor for dynorphin-(1-13)[178] in mouse vas deferens has been forwarded by Wüster et al.[205]

Tolerance without concomitant dependence[206] can be developed in this isolated organ by chronic in vivo administration of opioid agonists of different character.[205-208]

Characteristic feature is the lack (or low degree) of cross-tolerance between "μ" and "δ" agonists.[206,207] When high degree of tolerance had developed to both "μ" and "δ" agonists in the same preparation, dynorphin-(1-13) was capable of exerting almost full agonist effect.[205]

3. Rat Vas Deferens Preparation (RVD)

The innervation pattern of this isolated organ is similar to that of mouse vas deferens.[209] The mechanisms, controlling the release of neurotransmitter share also common features[195,210] save for the opioid mechanisms.[88]

The preparation has been found to be insensitive to the agonist action of morphine,[88] while it was susceptible to the inhibitory action of β-endorphin,[125,126] some related peptides[208,211,212] and a pentapeptide fragment (Phe-Phe-Gly-Leu-Met-NH$_2$) of substance P.[213] The inhibitory actions of these agents could be specifically antagonized by naloxone or naltrexone. Human[126,212] and porcine[214] β-EP have been found to be equipotent in guinea pig ileum, mouse and rat vas deferens preparations, while sheep β-EP and its D-Ala^2 derivative were moderately less potent opioid agonists in rat vas deferens than in guinea pig ileum.[211] However, (D-Ala^2, MePhe^4)-β$_s$-EP and (D-Ala^2,MePhe^4,Met(O)^5)-β$_s$-EP proved to be considerably more effective in rat vas deferens than in guinea pig ileum.[211] Some potent enkephalin analogs like FK 33-824,[212] D-Met, Pro-EA[212,214] (see also Chapter 7, Volume III) and D-Ala^2,Leu^5-enkephalin[125,212] (but not D-Ala^2,D-Leu^5-enkephalin) as well as the nonpeptide opioid compound etorphine[212] displayed also remarkable agonist efficacies not matching, though, their agonist potencies determined in other isolated organs.

Opioid peptides related structurally to β-LPH with shorter chain length than LPH(61-78) are practically devoid of agonist activity in this preparation[125,126,214] (see also Chapter 5, Volume III). The K_e values of classical opiate antagonists determined against either β-EP or D-Met, Pro-EA in isolated rat vas deferens[127,214-216] are similar to or only moderately higher than the figures obtained for the very same compounds in guinea pig ileum preparation. Furthermore, morphine[215,216] which has no agonist activity, retains some affinity to the opioid receptors in rat vas deferens.

We have found[127] that in vas deferens taken from young (10 to 16 days old) rats normorphine had well-defined opioid agonist action; during the maturation, this agonist property was gradually lost.

The opioid receptor type present in adult rat vas deferens was designated as "ε" for its selectivity to β-EP.[212]

4. Cat Nictitating Membrane Preparation (CNM)

The cat nictitating membrane receives noradrenergic motor innervation; the synaptic cleft, similarly to the morphological pattern in vas deferens, is rather narrow.[138,217,218] The release of norepinephrine, in addition to presynaptic α("α$_2$")-adrenergic receptors[219] has been reported to be controlled also by presynaptically located specific dopamine receptors.[136,143]

It was revealed by Trendelenburg[137] in 1957 that morphine inhibited the contractions of the nictitating membrane induced by nerve stimulation in vivo. This specific effect of morphine has been reproduced subsequently also in in vitro preparations.[120,139-141] This inhibitory action of alkaloid appears to involve an inhibition of norepinephrine release.[120,140]

Opioid peptides are also capable of depressing specifically the contractions elicited by nerve stimulation both in vitro[135,136,142,143,214] and in vivo.[144] Comparing the agonist potencies of a number of morphine congeners as well as of opioid peptides determined in cat nictitating membrane and longitudinal muscle-myenteric plexus of guinea pig ileum in vitro, the order of agonist potencies show a fair agreement in the two isolated organs (Figure 1). Furthermore, the very same holds for the K_e values of opiate antagonists and mixed agonist-antagonists, if determined against normorphine in these preparations (Figure 1). The values for some azidomorphine derivatives,[220] however, cannot be fitted into this correlation.

The K_e values of naltrexone determined against Met-E and β-EP in isolated cat nic-

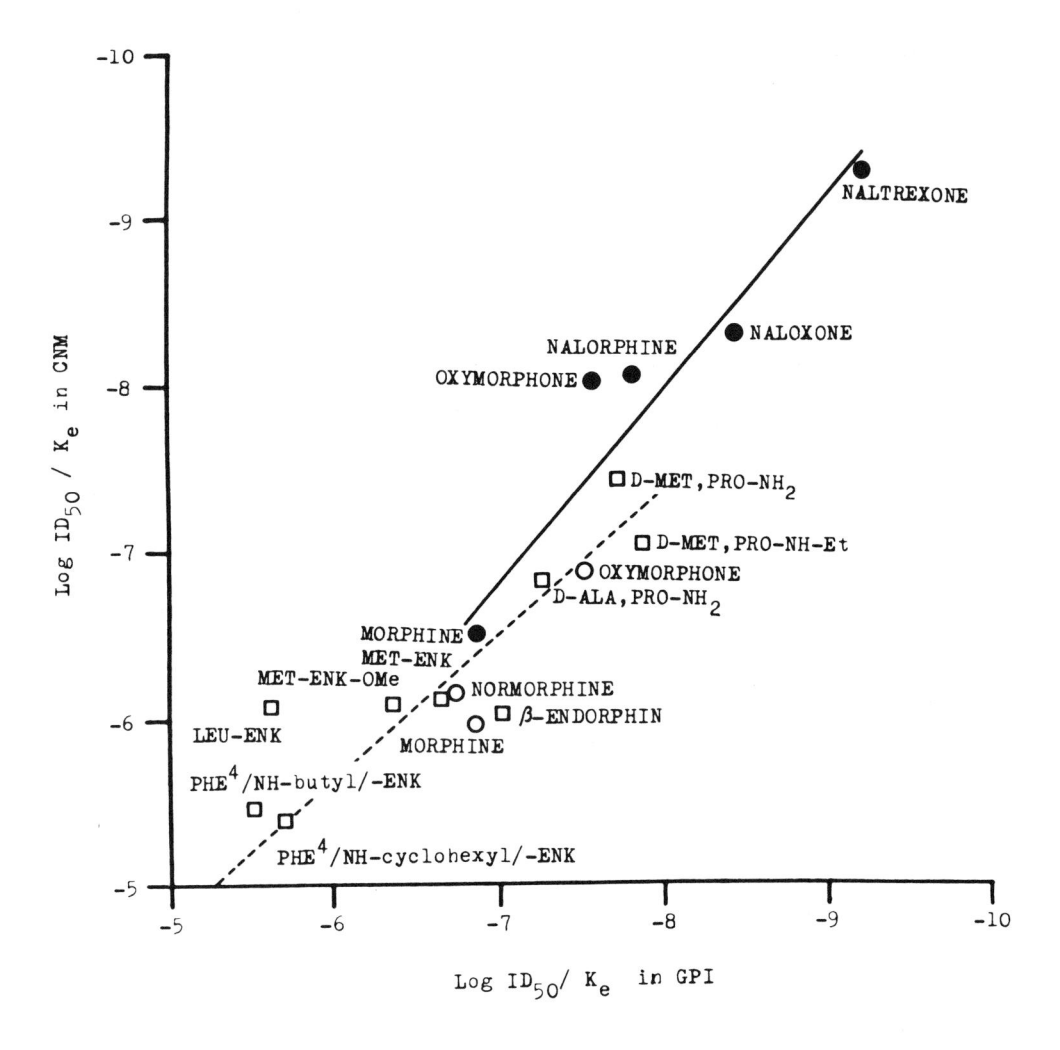

FIGURE 1. The logarithm of the ID$_{50}$ values of agonists (open symbols) and the K$_e$ values of antagonists and mixed antagonist-agonist (dark symbols) are plotted. Abbreviations: the enkephalin analgoues are specified by giving the substituents at position 2 and/or 5, referring to the structure of Met-enkephalin. Where other site is modified, it is specified accordingly.

For morphine:	ID$_{50}$	=	141.2 nM
	K$_e$	=	138.7 nM
oxymorphone:	ID$_{50}$	=	31.9 nM
	K$_e$	=	28.2 nM
nalorphine:	K$_e$	=	16.1 nM
naloxone:	K$_e$	=	3.9 nM

The correlations for both the agonist (correlation coefficient = 0.8853, standard error of the estimate = 0.3083, significance of regression (F ratio) = 0.0073%) and the antagonist (correlation coefficient = 0.9505, standard error of the estimate = 0.3569, significance of regression (F ratio) = 1.1578%) effects are significant.

titating membrane are significantly higher than the value obtained against normorphine[135,136,214] (see also Chapter 7, Volume III) (Table 1). Thus, in relation of normorphine-Met-E-β-EP, the order of agonist potencies in nictitating membrane bears similarity to the pattern found in guinea pig ileum, while their antagonizabilities by naltrexone resemble more closely to mouse vas deferens (see Figure 1, and also Chapter 7, Volume III, Table 1). It should be mentioned, however, that also in guinea pig ileum longitudinal muscle-myenteric plexus preparation, the tendency towards obtaining slightly higher K_e values for opiate antagonists if determined against some opioid peptides as compared to the values measured against normorphine, can be traced[175,214] (see also Chapter 7, Volume III) though in most cases the difference does not attain statistical significance.

5. Rabbit Ear Artery Preparation (ART)

The synaptic cleft in this noradrenergically innervated organ is considerably wider than in the case of vas deferens or the cat nictitating membrane.[218] In addition to the feedback modulation through presynaptic α-adrenoceptors, which is characteristic of adrenergic neurones, the release of norepinephrine is controlled by presynaptically located specific dopamine[136,221,222] and muscarinic cholinergic[223] receptors, and also by PGE_1.[224]

The constrictions of the isolated, perfused preparation[225] elicited by nerve stimulation have been found to be unaffected by normorphine.[134,136] Met- and Leu-E[134-136]

D-Met,[2]Nle[5]-enkephalin and Met-E-OMe[134] exerted, however, presynaptic inhibitory action in perfused, electrically stimulated rabbit ear artery in vitro in the concentration range comparable to that had been found in guinea pig ileum. With the exception of Met-E-OMe they were considerably weaker presynaptic inhibitors in the artery than in mouse vas deferens. Very high concentrations of opiate antagonists were necessary to antagonize the inhibitory effects of these peptides in electrically stimulated rabbit ear artery.[134,136] β-EP and D-Met, Pro-EA also depressed the constrictions of the organ elicited by nerve stimulation, acting presynaptically[135,136] (see also Chapter 7, Volume III, Table 1); their inhibitory potencies, however, were found to be much weaker in the ear artery than either in guinea pig ileum or mouse vas deferens.

The poor antagonizability of the inhibitory actions or by naltrexone/naloxone in the artery, was characteristic also of these peptides. No kinetic analysis of the antagonism has been performed to date; of the opioid peptides hitherto studied by our group,[136] the inhibitory effect of D-Met[2],Nle[5]-enkephalin could be reversed most readily by naloxone/naltrexone in this isolated organ. The inhibitory effect of Met-E but not that of the other opioid peptides could be counteracted with the dopamine antagonist sulpiride.[136]

Taking together, the presynaptic receptors in rabbit ear artery where these peptides exert their inhibitory action, bear some resemblance to the peptide-sensitive opioid receptors in mouse vas deferens.[136,214] This similarity is suggested by the order of inhibitory potencies, which, for the opioid substances studied by us was found to be D-Met[2],Nle[5]-enkephalin > Met-E > β-EP > D-Met,Pro-EA ≫ normorphine (ineffective) for the ear artery and D-Met[2],Nle[5]-enkephalin > Met-E > D-Met,Pro-EA > β-EP > normorphine for the mouse vas deferens.[135,136] The order of inhibitory potencies contrasts rather sharply the pattern that has been described either in guinea pig ileum or in cat nictitating membrane[135,136] (see also Figure 1).

V. BIOCHEMICAL METHODS

The biochemical methodology applied in opiate research should be divided into several branches. This covers

1. The detection and identification of active endogenous opioid substances in tissues/body fluids
2. The detection and identification of precursor(s) to these active substances and the clarification of processing of active components

3. The examination of biotransformation/biodegradation of active components
4. The characterization of biochemical events related to the action of opoid substances.

Topics 3 and 4 shall be treated in Chapter 4, Volume III and Chapter 1, Volume II though certain points related to paragraphs 1 and 2 are touched also in Chapters 2 and 4, Volume III. Some aspects, with special emphasis on the methodological pitfalls and also the useful innovations, will be surveyed briefly below.

The family of peptide sequences of natural origin, possessing opiate activity consists of several, structurally closely related members, which are contained by the very same endogenous peptide.[226-229] Thus, β-LPH[230] has been regarded for a while as a natural source of Met-E,[231] α-,[232] β-,[233] γ-[226] endorphins and the C' fragment[227] of β-LPH.[226-228] Enzymes which can form γ-endorphin, α-endorphin and even Met-E[234,235] from β-EP/β-LPH have been demonstrated to be present in brain tissue. Furthermore, it has been shown that porcine adenohypophysis contains endopeptidases to split Arg(60)-Tyr(61),[236] Leu(77)-Phe(78)[237] and Lys(79)-Asn(80)[236] peptide bonds of β-LPH.

During the extraction procedure, however, the physiological compartmentation of tissues is abolished, thus, the enzymes which normally may not have access to the endogenous substances in question, may carve peptide sequences during the extraction that might not be generated in the tissues under physiological conditions.[238-240]

If the tissue enzymes had been inactivated prior to the extraction by boiling or microwave irradiation[238] α-endorphin could no longer be detected in brain extract. It should be mentioned that freezing of tissue samples may result in the disruption of the cells and the release of lysosomal proteinases which may degrade the peptides before the heat inactivation of the enzymes.[239-241] Furthermore, since the enzyme that generates γ-endorphin from β-EP has been identified as lysosomal cathepsin D,[242] it is likely that the major if not the total amount of α- and γ-endorphins isolated from brain tissue may have been formed during the extraction-isolation procedure. However, as it was noted by Gráf et al.,[242] artifact formation mediated by cathepsin D must not divert the attention from the possibility that the enzyme, released from the cells under some physiological or pathological conditions might have an extracellular function in either inactivating β-EP or converting it into peptide(s) of physiological relevance.[243] As it was pointed out also for Met-E[229,244,245] (see also below) β-LPH/β-EP cannot be the natural precursor for the pentapeptide either.

Thus, the inactivation of the brain peptidases prior to extraction is a prerequisite in order to obtain reproducible results. This factor may explain at least in part, the conflicting results on the β-EP levels in the brain following hypophysectomy. Ogawa et al.,[246] sacrificing the animals by microwave irradiation, found the hypothalamic and midbrain levels of β-EP markedly reduced following hypophysectomy in rats, in contrast to a previous report.[247]

By all probabilities, rapid degradation by tissue enzymes may contribute to the fact that only minute amounts of α-neo-endorphin[248] and dynorphin-(1-13)[249] were detected in hypothalamic and pituitary extracts. The determination of the structure had to be carried out from nanomolar quantities of respective peptides. The clarification of the amino acid sequence of dynorphin-(1-13) was facilitated by the introduction of newly devised "spinning cup" technique.[249]

The common precursor to ACTH/β-LPH, the intermediates and end-products of processing have been demonstrated first by Mains et al.[250,251] by the combination of sequential, double-antibody immunoprecipitation procedure with antisera to endorphins and to ACTH, with the pulse labeling-chase technique, using also improved peptide-analytical methods (see also Chapter 4, Volume III). The experiments have been carried out in mouse pituitary tumor cell line /AtT-20/D-16v/ cultures; these cells release the precursor, the intermediates as well as the end-products of biosynthesis into the incubation medium.

Another approach was the use of a cell-free protein-synthesizing system (prepared from reticulocyte lysates) for examining the initial, direct product of translation of ACTH/β-LPH precursor and mRNA from anterior lobe of bovine pituitaries of AtT-20 cells.[252,253] The advantage of this method is that the signal peptide region of the precursor could also be characterized.[253]

A further method, involving the introduction of some recently developed techniques like DNA cloning and nucleotide sequence analysis into the opiate research[254,255] made possible to clarify the complete amino acid sequence of bovine ACTH/β-LPH precursor. The amino acid sequence corresponding to the nucleotide sequence of cloned DNA complementory to the mRNA coding for the precursor protein gave informations also on the primary structure of putative active sequences (e.g., "γ-MSH", "calcitonin-like structure") and "16 K" fragment[251] (which is a regular component among the end-products of processing)[256] the biological significance of which is unknown at the moment.

A common precursor of Met/Leu-E (an about 50,000 dalton protein) has been described in bovine adrenal medulla.[229]

Among the possible end-products of processing, claimed to occur naturally,[229] besides the enkephalins only Met-E-Arg^6-Phe^7 heptapeptide has been reported to possess significant opiate activity.[257]

REFERENCES

1. Berson, S. A. and Yalow, R. S., Quantitative aspects of the reaction between insulin and insulin-binding antibody, *J. Clin. Invest.*, 38, 1996, 1959.
2. Yalow, R. S., Glick, S. M., Roth, J., and Berson, S. A., Radioimmunoassay of human plasma ACTH, *J. Clin. Endocrinol. Metab.*, 24, 1219, 1964.
3. Paton, W. D. M. and Rang, H. P., The uptake of atropine and related drugs by intestinal smooth muscle of the guinea pig in relation to acetylcholine receptors, *Proc. Roy. Soc. London Ser. B.*, 163, 1, 1965.
4. Roth, J., Peptide hormone binding to receptors: a review of direct studies in vitro, *Metabolism*, 22, 1059, 1973.
5. Goldstein, A., Lowney, L., and Pal, B., Stereospecific and nonspecific interactions of the morphine congener levorphanol in subcellular fractions of the mouse brain, *Proc. Natl. Acad. Sci. USA*, 68, 1742, 1971.
6. Pert, C. B. and Snyder, S. H., Opiate receptor demonstration in nervous tissue, *Science*, 179, 1011, 1973.
7. Simon, E. J., Hiller, J. M., and Edelman, I., Stereospecific binding of the potent narcotic analgesic ³H-etorphine to rat brain homogenate, *Proc. Natl. Acad. Sci. USA*, 70, 1947, 1973.
8. Terenius, L., Characteristics of the "receptor" for narcotic analgesics in synaptic plasma membrane fractions from rat brain, *Acta Pharmacol. Toxicol.*, 33, 377, 1973.
9. Simantov, R. and Snyder, S. H., Brain-pituitary opiate mechanisms: pituitary opiate receptor binding, radioimmunoassays for methionine enkephalin and leucine-enkephalin and ³H-enkephalin interactions with the opiate receptor, in *Opiates and Endogenous Opioid Peptides*, Kosterlitz, H. W., Ed., Elsevier/North-Holland Biomedical Press, Amsterdam, 1976, 41.
10. Lefkowitz, R. J., Isolated hormone receptors, physiologic and clinical implications, *N. Engl. J. Med.*, 288, 1061, 1973.
11. Cuatrecasas, P., Membrane receptors, *Annu. Rev. Biochem.*, 43, 169, 1974.

12. Cuatrecasas, P., Criteria for and pitfalls in the identification of receptors, in *Pre- and Postsynaptic Receptors,* Usdin, E. and Bunney, W. E. Jr., Eds., Marcel Dekker, New York, 1975, 245.
13. Posner, B. I., Polypeptide hormone receptors: characteristics and applications, *Canad. J. Physiol. Pharmacol.,* 53, 689, 1975.
14. Boeynaems, J. M. and Dumont, J. E., Quantitative analysis of the binding of ligands to their receptors, *J. Cyclic Nucleotide Res.,* 1, 123, 1975.
15. Höllt, V., The opiate receptors, in *Developments in Opiate Research,* Herz, A., Ed., Marcel Dekker, New York, 1978, 1.
16. O'Brien, R. D., Ed., A comprehensive treatise. General principles and procedures, *The Receptors,* Vol. 1, Plenum Press, New York, 1979.
17. Leslie, F. M., Chavkin, V. and Cox, B. M., Ligand specificity of opioid binding sites in brain and peripheral tissues, in *Endogenous and Exogenous Opiate Agonists and Antagonists,* Way, E. L., Ed., Pergamon Press, New York, 1979, 109.
18. Snyder, S. H., Opiate receptors and internal opiates, *Scientific American,* 236, 44, 1977.
19. Beckett, A. H. and Casy, A. F., Synthetic analgesics: stereochemical considerations, *J. Pharm. Pharmacol.,* 6, 986, 1954.
20. Snyder, S. H. and Pert, C. B., Membrane receptor, in *Opiate Receptor Mechanisms,* Snyder, S. H. and Matthysse, S., Eds., The MIT Press, Massachusetts, 1975, 26.
21. Creese, I. and Snyder, S. H., Receptor binding and pharmacological activity in the guinea pig intestine, *J. Pharmacol. Exp. Ther.,* 194, 205, 1975.
22. Lineweaver, H. and Burk, D., Determination of enzyme dissociation constants, *J. Am. Chem. Soc.,* 56, 658, 1934.
23. Scatchard, G., The attraction of proteins for small molecules and ions, *N.Y. Acad. Sci.,* 51, 660, 1949.
24. Hill, A. V., The possible effect of the aggregation of haemoglobin on its dissociation curves, *J. Physiol.,* 40, 190, 1910.
25. Baulieu, E. E. and Raynaud, J. P., A "Proportion graph" method for measuring binding systems, *Eur. J. Biochem.,* 13, 293, 1970.
26. Thompson, C. J. and Klotz, I. M., Macromolecule — small molecule interaction: analytical graphical reexamination, *Arch. Biochem. Biophys.,* 147, 178, 1971.
27. Fischel, S. V. and Medzihradsky, F., Scatchard analysis of opiate receptor binding, in *Endogenous and Exogenous Opiate Agonists and Antagonists,* Way, E. L., Ed., Pergamon Press, New York, 1979, 91.
28. Henis, Y. I. and Levitzki, A., An analysis on the slope of Scatchard plots, *Eur. J. Biochem.,* 71, 529, 1976.
29. Kosterlitz, H. W., Lord, J. A. H., Paterson, S. J., and Waterfield, A. A., Effects of changes in the structure of enkephalins and of narcotic analgesic drugs on their interactions with μ- and δ-receptors, *Br. J. Pharmacol.,* 68, 333, 1980.
30. Cheng, Y. C. and Prusoff, W. H., Relationship between the inhibition constant (K_i) and the concentration of inhibitor (IC_{50}) of an enzymatic reaction, *Biochem. Pharmacol.,* 22, 3099, 1973.
31. Martin, W. R., Eades, C. G., Thompson, J. A., Huppler, R. E. and Gilbert, P. E., The effects of morphine- and nalorphine-like drugs in the nondependent and morphine-dependent chronic spinal dog, *J. Pharmacol. Exp. Ther.,* 197, 517, 1976.
32. Lord, J. A. H., Waterfield, A. A., Hughes, J., and Kosterlitz, H. W., Endogenous opioid peptides: multiple agonists and receptors, *Nature (London),* 267, 495, 1977.
33. Wüster, M., Schulz, R., and Herz, A., specificity of opioids towards the μ-, δ- and ϵ-opiate receptors, *Neurosci. Lett.,* 15, 193, 1979.
34. Wüster, M., Schulz, R., and Herz, A., Highly specific opiate receptors for dynorphin-(1-13) in the mouse vas deferens, *Eur. J. Pharmacol.,* 62, 235, 1980.
35. Chang, K.-J., Miller, R. J., and Cuatrecasas, P., Interaction of enkephalin with opiate receptors in intact cultured cells, *Mol. Pharmacol.,* 14, 961, 1978.
36. Chang, K.-J., Cooper, B. R., Hazum, E., and Cuatrecasas, P., Multiple opiate receptors: different regional distribution in the brain and differential binding of opiates and opioid peptides, *Mol. Pharmacol.,* 16, 91, 1979.
37. Zukin, R. S., Walczak, S., and Makman, M. H., GTP modulation of opiate receptors in regions of rat brain and possible mechanism of GTP action, *Brain Res.,* 186, 238, 1980.
38. Pert, C. B. and Taylor, D., Type 1 and type 2 opiate receptors: a subclassification scheme based upon GTP's differential effects on binding, in *Endogenous and Exogenous Opiate Agonists and Antagonists,* Way, E. L., Ed., Pergamon Press, New York, 1979, 87.
39. Smith, A. P. and Loh, H. H., Heterogeneity of opiate-receptor interaction, *Pharmacology,* 20, 57, 1980.

40. Snyder, S. H. and Goodman, R. R., Multiple neurotransmitter receptors, *J. Neurochem.*, 35, 5, 1980.
41. Atweh, S. F. and Kuhar, M. J., Autoradiographic localization of opiate receptors in rat brain. I. Spinal cord and lower medulla, *Brain Res.*, 124, 53, 1977.
42. Kream, R. M. and Zukin, R. S., Binding characteristics of a potent enkephalin analog, *Biochem. Biophys. Res. Commun.*, 90, 99, 1979.
43. Akil, H., Hewlett, W., Barchas, J. D., and Li, C. H., Characterization of ³H-β-endorphin binding in rat brain, in *Endogenous and Exogenous Opiate Agonists and Antagonists*, Way, E. L., Ed., Pergamon Press, New York, 1979, 123.
44. Simon, E. J. and Hiller, J. M., The opiate receptors, *Annu. Rev. Pharmacol. Toxicol.*, 18, 371, 1978.
45. Pert, C. B. and Snyder, S. H., Opiate receptor binding of agonists and antagonists affected differentially by sodium, *Mol. Pharmacol.*, 10, 868, 1974.
46. Berson, S. A. and Yalow, R. S., Quantitative aspects of the reaction between insulin and insulin-binding antibody, *J. Clin. Invest.*, 38, 1996, 1959.
47. Yalow, R. S., Glick, S. M., Roth, J., and Berson, S. A., Radioimmunoassay of human plasma ACTH, *J. Clin. Endocrinol. Metab.*, 24, 1219, 1964.
48. Hughes, J., Smith, T. W., Kosterlitz, H. W., Fothergill, L. A., Morgan, B. A., and Morris, H. R., Identification of two related pentapeptides from the brain with potent opiate agonist activity, *Nature (London)*, 258, 577, 1975.
49. Guillemin, R., Ling, N., and Burgus, R., Endorphines, peptides d'origine hypothalamique et neurohypophysaire a activite morphinomimetique. Isolement et structure moleculaire de l' α-endorphine, *C. R. Acad. Sci. (Paris)*, 282, 783, 1976.
50. Li, C. H. and Chung, D., Isolation and structure of an untriakontapeptide with opiate activity from camel pituitary glands, *Proc. Natl. Acad. Sci. USA*, 73, 1145, 1976.
51. Simantov, R. and Snyder, S. H., Brain-pituitary opiate mechanisms: pituitary opiate receptor binding, radio-immunoassays for methionine enkephalin and leucine enkephalin, and ³H-enkephalin interactions with the opiate receptor, in, *Opiates and Endogenous Opioid Peptides*, Kosterlitz, H. W., Ed., Elsevier/North-Holland Biomedical Press, Amsterdam, 1976, 41.
52. Guillemin, R., Ling, N., and Vargo, J., Radioimmunoassays for α-endorphin and β-endorphin, *Biochem. Biophys. Res. Commun.*, 77, 361, 1977.
53. Borvendég, J., Gráf, L., Hermann, I., Palkovits, M., and Merétey, K., Radioimmunoassay of β-endorphin: immunoreactive substances in the brain and the pituitary, in *Endorphins '78*, Gráf, L., Palkovits, M., and Rónai, A. Z., Eds., Publishing House of the Hungarian Academy of Science, Budapest, 1978, 177.
54. Höllt, V., Przewlocki, R., and Herz, A. Radioimmunoassay of β-endorphin. Basal and stimulated levels in extracted rat plasma, *Naunyn-Schmiedeb. Arch. Pharmacol.* 303, 171, 1978.
55. Wesche D., Höllt, V., and Herz, A., Radioimmunoassay of enkephalins. Regional distribution in rat brain after morphine treatment and hypophysectomy, *Naunyn-Schmiedeb. Arch. Pharmacol.*, 301, 79, 1977.
56. Gros, C., Pradelles, P., Rouget, C., Bepoldin, O., Dray, F., Fournie-Zaluski, M. C., Roques, B. P., Pollard, H., Llorens-Cortes, C., and Schwartz, J. C., Radioimmunoassay of methionine- and leucine-enkephalins in regions of rat brain and comparison with endorphins estimated by radioreceptor assay, *J. Neurochem.*, 31, 29, 1978.
57. Rossier, J., Bayon, A., Vargo, T. M., Ling, N., Guillemin, R., and Bloom, F., Radioimmunoassay of brain peptides: evaluation of methodology for the assay of β-endorphin and enkephalin, *Life Sci.*, 21, 847, 1977.
58. Rossier, J., Vargo, T. M., Minick, S., Ling, N., Bloom, F. E., and Guillemin, R., Regional dissociation of β-endorphin and enkephalin contents in rat brain and pituitary, *Proc. Natl. Acad. Sci. USA*, 74, 5162, 1977.
59. Llorens-Cortes, C., Pollard, H., Schwartz, J. C., Pradelles, P., Gros, C., and Dray, F., Endorphins in several regions of rat brain: large differences between radioimmunoassay and radioreceptor assay, *Eur. J. Pharmacol.*, 46, 73, 1977.
60. Hong, J. S., Yang, H.-Y., Fratta, W., and Costa, E., Determination of methionine-enkephalin in discrete regions of rat brain, *Brain Res.*, 134, 383, 1977.
61. Yang, H.-Y., Hong, J. S., and Costa, E., Regional distribution of leu- and met-enkephalin in rat brain, *Neuropharmacology*, 16, 303, 1977.
62. Kobayashi, R. M., Palkovits, M., Miller, R. J., Chang, K.-J., and Cuatrecasas, P., Distribution of enkephalin in the brain is unaltered by hypophysectomy, *Life Sci.*, 22, 527, 1978.
63. Krieger, D. T., Liotta, A., Suda, T., Palkovits, M., and Brownstein, M. J., Presence of immunoassayable β-lipotropin in bovine brain and spinal cord: lack of concordance with ACTH concentrations, *Biochem. Biophys. Res. Commun.*, 76, 930, 1977.

64. Krieger, D. T., Liotta, A., and Li, C. H., Human plasma immunoreactive β-lipotropin: correlation with basal and stimulated plasma ACTH concentrations, *Life Sci.*, 21, 1771, 1977.
65. Krieger, D. T., Plasma lipotropin and endorphin in the human, in *Endorphins '78*, Gráf, L., Palkovits, M., and Rónai, A. Z., Eds., Publishing House of the Hungarian Academy of Science, Budapest, 1978, 275.
66. Wiedemann, E., Saito, T., Linfoot, J. A., and Li, C. H., Radioimmunoassay of human β-lipotropin in unextracted plasma, *J. Clin. Endocrinol. Metab.*, 45, 1108, 1977.
67. Gilkes, J. J. H., Reed, L. H., and Besser, G. M., Plasma immunoreactive corticotrophin and lipotrophin in Cushing's syndrome and Addison's disease, *Br. Med. J.*, 1, 996, 177.
68. Tanaka, K., Nicholson, W. E., and Orth, D. N., The nature of the immunoreactive lipotropins in human plasma and tissue extract, *J. Clin. Invest.*, 62, 94, 1978.
69. Chang, W.-C., Rao, A. J., and Li, C. H., Rate of disappearance of human lipotropin and endorphin in adult male rats as estimated by radioimmunoassay, *Int. J. Peptide Protein Res.*, 11, 93, 1978.
70. Ross, M., Ghazarossian, V., Cox, B. M., and Goldstein, A., Radioimmunoassays for endorphin: comparison of properties of two antisera, *Life Sci.*, 22, 1123, 1978.
71. Mains, R. E. and Eipper, B. A., Studies on the common precursor to ACTH and endorphin, in *Endorphins '78*, Gráf, L., Palkovits, M., and Rónai, A. Z., Eds., Publishing House of the Hungarian Academy of Science, Budapest, 1978, 79.
72. Rossier, J., Audigier, Y., Ling, N., Cros, J., and Udenfriend, S., Met-enkephalin-Arg^6-Phe^7, present in high amounts in brain of rat, cattle and man, is an opioid agonist, *Nature (London)*, 288, 88, 1980.
73. Lewis, R. V., Stern, A. S., Kimura, S., Rossier, J., Stein, S., and Udenfriend, S., An about 50,000-dalton protein in adrenal medulla: a common precursor of /Met/- and /Leu/-enkephalin, *Science*, 208, 1459, 1980.
74. Rossier, J., Dean, D., Livett, B., and Udenfriend, S., Bovine adrenal cells release enkephalins and several characterized putative enkephalin precursors, *Neurosci. Lett.*, Suppl. 5, 358, 1980.
75. Gráf, L., Hollósi, M., Barna, I., Hermann, I., Borvendég, J., and Ling, N., Probing the biologically and immunologically active conformation of β-endorphin: studies on C-terminal deletion analogs, *Biochem. Biophys. Res. Commun.*, 95, 1623, 1980.
76. Li, C. H., Rao, A. J., Doneen, B. A., and Yamashiro, D., β-endorphin: lack of correlation between opiate activity and immunoreactivity by radioimmunoassay, *Biochem. Biophys. Res., Commun.*, 75, 576, 1977.
77. Li, C. H., β-endorphin: aspects of structure-activity relationship, in *Endorphins '78*, Gráf, L., Palkovits, M., and Rónai, A. Z., Eds., Publishing House of the Hungarian Academy of Science, Budapest, 1978, 15.
78. Mains, R. E., Eipper, B. A., and Ling, N., Common precursor to corticotropins and endorphins, *Proc. Natl. Acad. Sci. USA*, 74, 3014, 1977.
79. Smyth, D. G., Massey, D. E., Zakarian, S., and Finnie, M. D. A., Endorphins are stored in biologically active and inactive forms: isolation of α-N-acetyl peptides, *Nature (London)*, 279, 252, 1979.
80. Zakarian, S. and Smyth, D. G., Specific processing of endorphins in rat brain, in *Endogenous and Exogenous Opiate Agonists and Antagonists*, Way, E. L., Ed., Pergamon Press, New York, 1979, 301.
81. Seizinger, B. R. and Höllt, V., In vitro biosynthesis and N-acetylation of β-endorphin in pars intermedia of rat pituitary, *Biochem. Biophys. Res. Commun.*, 96, 535, 1980.
82. Zakarian, S. and Smyth, D. G., Distribution of active and inactive forms of endorphins in rat pituitary and brain, *Proc. Natl. Acad. Sci. USA*, 76, 5972, 1979.
83. Li, C. H., Chang, W.-C., Yamashiro, D., and Tseng, L.-T., β-endorphin: deletion of a simple amino acid residue abolishes immunoreactivity but retains opiate potency, *Biochem. Biophys. Res. Commun.*, 87, 693, 1979.
84. Jaffe, J. H., Narcotic analgesics, in *The Pharmacological Basis of Therapeutics*, Goodman, L. S. and Gilman, A., Eds., Macmillan, New York, 1970, 237.
85. Trendelenburg, P., Physiologische and pharmakologische Versuche über Dunndarmperistaltik, *Naunyn-Schmiedeb. Arch. Exp. Path. Pharmak.*, 81, 55, 1917.
86. Cox, B. M. and Weinstock, M., The effect of analgesic drugs on the release of acetylcholine from electrically stimulated guinea-pig ileum, *Br. J. Pharmacol.*, 27, 81, 1966.
87. Kosterlitz, H. W. and Watt, A. J., Kinetic parameters of narcotic agonists and antagonists, with particular reference to N-allylnoroxymorphone (naloxone), *Br. J. Pharmacol.*, 33, 266, 1968.
88. Hughes, J., Kosterlitz, H. W., and Leslie, F. M., Effect of morphine on adrenergic transmission in the mouse vas deferens. Assessment of agonist and antagonist potencies of narcotic analgesics, *Br. J. Pharmacol.*, 53, 371, 1975.
89. Kosterlitz, H. W. and Waterfield, A. A., In vitro models in the study of structure-activity relationships of narcotic analgesics, *Annu. Rev. Pharmacol.*, 15, 28, 1975.

90. Eddy, N. B. and Leimbach, D., Synthetic analgesics. II. Dithienylbutenyl- and dithienylbutylamines, *J. Pharmac. Exp. Ther.*, 107, 385, 1953.
91. Archer, S., Albertson, N. F., Harris, L. S., Pierson, A. K., and Bird, J. G., Pentazocine. Strong analgesics and analgesic antagonists in the benzomorphan series, *J. Med. Chem.*, 7, 123, 1964.
92. Braude, M. C., Harris, L. S., May, E. L., Smith, J. P., and Villarreal, J. E., Eds., Advances in biochemical psychopharmacology, in *Narcotic Antagonists*, Raven Press, New York, 1974.
93. Hughes, J., Search for the endogenous ligand of the opiate receptor, in *Opiate Receptor Mechanisms*, Snyder, S. H. and Matthysse, S., Eds., MIT Press, Massachusetts, 1975, 55.
94. Hughes, J., Isolation of an endogenous compound from the brain with pharmacological properties similar to morphine, *Brain Res.*, 88, 295, 1975.
95. Hughes, J., Smith, T. W., Kosterlitz, H. W., Forthergill, L. A., Morgan, B. A., and Morris, H. R., Identification of two related pentapeptides from the brain with potent opiate agonist activity, *Nature (London)*, 258, 577, 1975.
96. Hirst, G. D. S. and McKirdy, H. C., Synaptic potentials recorded from neurones of the submucosus plexus of guinea pig small intestine, *J. Physiol.*, 249, 369, 1975.
97. Schultzberg, M., Dreyfus, C. F., Gershon, M. D., Hökfelt, T., Elde, R., Nelsson, G., Said, S., and Goldstein, M., VIP-, enkephalin-, substance P-, and somatostatin-like immunoreactivity in neurons intrinsic to the intestine: immunohistochemical evidence from organotypic tissue culture, *Brain Res.*, 155, 239, 1978.
98. Paton, W. D. M., The action of morphine and related substances on contraction and acetylcholine output of coaxially stimulated guinea pig ileum, *Br. J. Pharmacol.* 12, 119, 1957.
99. Schultz, R., The use of isolated organs to study the mechanism of action of narcotic analgesics, in *Developments in Opiate Research*, Herz, A., Ed., Marcel Dekker, New York, 1978, 241.
100. Puig, M. M., Gascon, P., Craviso, G. L., and Musacchio, J. M., Electrically induced release of an endogenous opiate receptor ligand in the guinea pig ileum, *Science*, 145, 419, 1977.
101. Schluz, R., Wüster, M., Simantov, R., Snyder, S. H., and Herz, A., Electrically stimulated release of opiate-like material from the myenteric plexus of the guinea pig ileum, *Eur. J. Pharmacol.*, 41, 347, 1977.
102. Burks, T. F., Mediation by 5-hydroxytryptamine of morphine stimulant actions in dog intestine, *J. Pharmacol. Exp. Ther.*, 185, 530, 1973.
103. Nijkamp, F. P. and van Ree, J. M., Effects of endorphins on different parts of the gastrointestinal tract of rat and guinea pig in vitro, *Br. J. Pharmacol.*, 68, 599, 1980.
104. Schaumann, O., Jochum, K., and Schmidt, H., Analgetika und Darmmotorik. II. Zum Mechanismus der Peristaltik, *Naunyn-Schmiedeb. Arch. Exp. Path. Pharmak.*, 219, 302, 1953.
105. Kosterlitz, H. W. and Robinson, J. A., Inhibition of peristaltic reflex of the isolated guinea-pig ileum, *J. Physiol. (London)*, 136, 249, 1957.
106. Schaumann, W., The paralysing action of morphine on the guinea-pig ileum, *Br. J. Pharmacol.*, 10, 456, 1955.
107. Gaddum, J. H. and Piccarelli, Z. P., Two kinds of tryptamine receptor, *Br. J. Pharmacol.*, 12, 323, 1957.
108. Lewis, G. P., The inhibition by morphine on the action of smooth muscle stimulants on the guinea pig intestine, *Br. J. Pharmacol.*, 15, 425, 1960.
109. Vogt, W., Antagonismus von Atropin und Morphin gegenüber der darmerregenden Wirkung von Darmstoff, *Naunyn-Schmiedeb. Arch. Exp. Path. Pharmak.*, 235, 550, 1959.
110. Ross, C. A., Ludden, C. T., and Stone, C. A., Action of angiotensin on isolated guinea pig ileum, *Proc. Soc. Exp. Biol. Med.*, 105, 558, 1960.
111. Zsêli, J., Vizi, E. S., and Knoll, J., Az azidomorphin és a morphin intestinális hatásának in vitro tanulmányozása, *Kiserletes Orvostudomany*, 28, 202, 1976.
112. Jaques, R., Suppression by morphine and other analgesic compounds of the smooth muscle contraction produced by arachidonic acid peroxide. An in vitro method for detecting potential analgesics, *Helv. Physiol. Acta*, 23, 156, 1965.
113. Harry, J., The action of drugs on the circular muscle strip from the guinea pig isolated ileum, *Br. J. Pharmacol.*, 20, 399, 1963.
114. Lees, G. M., Kosterlitz, H. W., and Waterfield, A. A., Characteristics of morphine-sensitive release of neurotransmitter substances, in *Agonist and Antagonist Actions of Narcotic Analgesic Drugs*, Kosterlitz, H. W., Collier, H. O. J., and Villarreal, J. E., Eds., Macmillan, London, 1972, 142.
115. Kosterlitz, H. W., Waterfield, A. A., and Berthoud, V., Assessment of the agonist and antagonist properties of narcotic analgesic drugs by their actions on the morphine receptor in the guinea pig ileum, in *Narcotic Antagonists*, Braude, M. C., Harris, L. S., May, E. L., Smith, J. P., and Villarreal, J. E., Eds., Raven Press, New York, 1974, 319.
116. Lord, J. A. H., Waterfield, A. A., Hughes, J., and Kosterlitz, H. W., Multiple opiate receptors, in *Opiates and Endogenous Opioid Peptides*, Kosterlitz, H. W., Ed., Elsevier/North-Holland Biomedical Press, Amsterdam, 1976, 275.

117. Lord, J. A. H., Waterfield, A. A., Hughes, J., and Kosterlitz, H. W., Endogenous opioid peptides: multiple agonists and receptors, *Nature (London)*, 267, 495, 1977.
118. Rónai, A. Z., Gráf, L., Székely, J. I., Dunai-Kovács, Z., and Bajusz, S., Differential behaviour of LPH-(61-91)-peptide in different model systems: comparison of the opioid activities of LPH-(61-91)-peptide and its fragments, *FEBS Lett.*, 74, 182, 1977.
119. Hughes, J., Kosterlitz, H. W., and Smith, T. W., The distribution of methionine-enkephalin and leucine-enkephalin in the brain and peripheral tissues, *Br. J. Pharmacol.*, 61, 639, 1977.
120. Henderson, G., Hughes, J., and Kosterlitz, H. W., The effects of morphine on the release of noradrenaline from the cat isolated nictitating membrane and the guinea pig ileum myenteric plexus-longitudinal muscle preparation, *Br. J. Pharmacol.*, 53, 505, 1975.
121. Shimo, Y. and Ishi, T., Effects of morphine on nonadrenergic inhibitory responses of the guinea pig taenia coli, *J. Pharm. Pharmacol.*, 30, 596, 1978.
122. Paton, W. D. M. and Vane, J. R., An analysis of the responses of the isolated stomach to electrical stimulation and to drugs, *J. Physiol. (London)*, 165, 10, 1963.
123. Szabolcsi, I., Vizi, E. S., and Knoll, J., *Symposium on Current Problems in the Pharmacology of Analgesics*, Vizi, E. S., Ed., Publishing House of the Hungarian Academy of Science, Budapest, 1974, 57.
124. Henderson, G., Hughes, J., and Kosterlitz, H. W., A new example of morphine-sensitive neuroeffector junction: adrenergic transmission in the mouse vas deferens, *Br. J. Pharmacol.*, 46, 764, 1972.
125. Lemaire, S., Magnan, J., and Regoli, D., Rat vas deferens: a specific bioassay for endogenous opioid peptides, *Br. J. Pharmacol.*, 64, 327, 1978.
126. Schulz, R., Faase, E., Wüster, M., and Herz, A., Selective receptors for β-endorphin on the rat vas deferens, *Life Sci.*, 24, 843, 1979.
127. Rónai, A. Z., Berzétei, I., and Kurgyis, J., Opioid effects in developing rat vas deferens, *J. Mol. Neurobiol.*, submitted.
128. Widy-Tyszkievicz, E., Luczak, A., Gumulka, W., and Szreniawski, Z., Antagonism by morphine and pethidine of the contractant effect of PGE$_2$ on the isolated uterus of the rat, *Naunyn-Schmiedeb. Arch. Pharmacol.*, Suppl. 294, R5, 1976.
129. Weinstock, M., Peripheral tissues, in *Narcotic Drugs, Biochemical Pharmacology. V. Sites of Action of Narcotic Analgesic Drugs*, Clouet, D. H., Ed., Plenum Press, New York, 1971, 394.
130. Macht, D. I., On the relation between the chemical structure of the opium alkaloids and their physiological action on smooth muscle with a pharmacological and therapeutic study of some benzyl esters I. On the relation of the chemical structure of the opium alkaloids to their action on smooth muscle, *J. Pharm. Exp. Ther.*, 11, 389, 1918.
131. Ockerblad, N. F., Carlson, H. E., and Simon, J. F., The effect of morphine upon the human ureter, *J. Urol.*, 33, 356, 1935.
132. Kosterlitz, H. W. and Taylor, D. W., The effects of morphine on the vagal inhibition of the heart, *Br. J. Pharmacol.*, 14, 209, 1959.
133. Montel, H. and Starke, K., Effects of narcotic analgesics and their antagonists on the rabbit isolated heart and its adrenergic nerves, *Br. J. Pharmacol.*, 49, 628, 1973.
134. Knoll, J., Neuronal peptide (enkephalin) receptors in the ear artery of the rabbit, *Eur. J. Pharmacol.*, 39, 403, 1976.
135. Rónai, A. Z., Berzétei, I., Székely, J. I., and Bajusz, S., The effect of synthetic and natural opioid peptides in isolated organs, in *Characteristics and Function of Opioids*, van Ree, J. M. and Terenius, L., Eds., Elsevier/North-Holland Biomedical Press, Amsterdam, 1978, 493.
136. Rónai, A. Z. and Berzétei, I., Similarities and differences of opioid receptors in different isolated organs, in *Endorphins '78*, Gráf, L., Paklovits, M., and Rónai, A. Z., Eds., Publishing House of the Hungarian Academy of Science, Budapest, 1978, 237.
137. Trendelenburg, U., The action of morphine on the superior cervical ganglion and on the nictitating membrane of the cat, *Br. J. Pharmacol.*, 12, 79, 1957.
138. Thompson, J. W., Studies on the responses of the isolated nictitating membrane of the cat, *J. Physiol. (London)*, 141, 46, 1958.
139. Thompson, J. W., *The Cat's Nictitating Membrane as an Isolated Preparation*, Ph.D. thesis, University of London, 1960.
140. Cairnie, A. B., Kosterlitz, H. W., and Taylor, D. W., Effect of morphine on some sympathetically innervated effectors, *Br. J. Pharmacol.*, 17, 539, 1961.
141. Enero, M. A., Properties of the peripheral opiate receptors in the cat nictitating membrane, *Eur. J. Pharmacol.*, 45, 349, 1977.
142. Knoll, J., Illés, P., and Medzihradszky, K., The action of enkephalins and enkephalin analogues on neurotransmission in the isolated nictitating membrane of the cat, *J. Pharm. Pharmacol.*, 30, 394, 1978.

143. Illés, P. and Knoll, J., Opiate and dopamine receptors in the isolated nictitating membrane of the cat, *Pol. J. Pharmacol. Pharm.*, 30, 293, 1978.
144. Yen. M.-H., Ku, P. Y., and Lee, H.-K., The effect of /D-Ala²/-Met-enkephalin on the contraction of nictitating membrane in cats, *Eur. J. Pharmacol.*, 63, 213, 1980.
145. Wouters, W. and van den Bercken, J., Depression of s-IPSP and hyperpolarization by Met-enkephalin in frog sympathetic ganglion, in *Characteristics and Function of Opioids*, van Ree, J. M. and Terenius, L., Eds., Elsevier/North-Holland Biomedical Press, Amsterdam, 1978, 111.
146. Wouters, W. and van den Bercken, J., Hyperpolarisation and depression of slow synaptic inhibition enkephalin in frog sympathetic ganglion, *Nature (London)*, 277, 534, 1979.
147. Wouters, W. and van den Bercken, J., Effects of Met-enkephalin on slow synaptic inhibition in frog sympathetic ganglion, *Neuropharmacology*, 19, 237, 1980.
148. Su, P. C., Gollapudi, L. M., and Rosen, A. D., Reduction of quantal release by methionine and leucine enkephalin on motor nerve terminals of mice, *Clin. Res.*, 26, 612 A, 1978.
149. Blumberg, H. and Dayton, H. B., Naloxone, naltrexone and related noroxymorphones, in *Narcotic Antagonists*, Braude, M. C., Harris, L. S., May, E. L., Smith, J. P., and Villarreal, J. E., Eds., Raven Press, New York, 1974, 33.
150. Arunlakshana, O. and Schild, H. O., Some quantitative uses of drug antagonists, *Br. J. Pharmacol.*, 14, 48, 1959.
151. Paton, W. D. M., A theory of drug action based on the rate of drug-receptor combination, *Proc. R. Soc. Biol.*, 154, 21, 1961.
152. Kosterlitz, H. W., Collier, H. O. J., and Villarreal, J. E., Eds., *Agonist and Antagonist Actions of Narcotic Analgesic Drugs*, Macmillan Press, London, 1972.
153. Martin, W. R., Eades, C. G., Thompson, J. A., Huppler, R. A., and Gilbert, P. E., The effects of morphine- and nalorphine-like drugs in the nondependent and morphine-dependent chronic spinal dog, *J. Pharm. Exp. Ther.*, 197, 517, 1976.
154. Gilbert, P. E. and Martin, W. R., The effects of morphine and nalorphine-like drugs in the nondependent, morphine-dependent and cyclazocine-dependent chronic spinal dog, *J. Pharm. Exp. Ther.*, 198, 66, 1976.
155. Bennett, T. and Cabb, J. L., Studies on the avian gizzard Auerbach's plexus, *Z. Zellforsch.*, 99, 109, 1969.
156. Jensen, K. R., Polak, J. M., van Noorden, S., Bloom, S. R., and Burnstock, G., Peptide-containing neurones connect the two ganglionated plexuses of the enteric nervous system, *Nature (London)*, 283, 391, 1980.
157. Gershon, M. D. and Altman, R. F., An analysis of the uptake of 5-hydroxytryptamine by the myenteric plexus of the small intestine of the guinea pig, *J. Pharm. Exp. Ther.*, 179, 29, 1971.
158. Gabella, G., Fine structure of the myenteric plexus in the guinea pig ileum, *J. Anat.*, 111, 69, 1972.
159. Bennett, M. R. and Rogers, D. C., A study on the innervation of the taenia coli, *J. Cell Biol.*, 33, 573, 1967.
160. Rogers, D. C. and Burnstock, G., The intestinal cell and its place in the concept of the autonomic ground plexus, *J. Comp. Neurol.*, 126, 255, 1966.
161. Paton, W. D. M., Vizi, E. S., and Zar, M. A., The mechanisms of acetylcholine release from parasympathetic nerves, *J. Physiol.*, 215, 819, 1971.
162. Vizi, E. S., The role of α-adrenoceptors situated in Auerbach's plexus in the inhibition of gastrointestinal motility, in *Physiology of Smooth Muscle*, Bülbring, E. and Shuba, M. F., Eds., Raven Press, New York, 1976, 357.
163. Burnstock, G., Cholinergic, adrenergic, and purinergic neuromuscular transmission, *Fed. Proc. Fed. Am. Soc. Exp. Biol.*, 36, 2434, 1977.
164. Furness, J. B. and Costa, M., The adrenergic innervation of the gastrointestinal tract, *Ergebn. D. Physiol.*, 69, 1974.
165. Ahlquist, R. P. and Levy, B., Adrenergic receptor mechanism of canine ileum, *J. Pharm. Exp. Ther.*, 127, 146, 1959.
166. Gershon, M. D., Inhibition of gastrointestinal movement by sympathetic nerve stimulation: the site of action, *J. Physiol.*, 189, 317, 1967.
167. Van Nueten, J. M., Janssen, P. A. J., and Fontaine, J., Unexpected reversal effects of naloxone on the guinea pig ileum, *Life Sci.*, 18, 803, 1976.
168. Kroemer, W. and Pretzlaff, W., In vitro evidence for the participation of intestinal opioids in the control of peristalsis in the guinea pig small intestine, *Naunyn-Schmiedeb. Arch. Pharmacol.*, 309, 153, 1979.
169. Kadlec, O., Masek, K., and Seferna, I., Modulation by prostaglandins of the release of acetylcholine and noradrenaline in guinea pig isolated ileum, *J. Pharm. Exp. Ther.*, 205, 635, 1978.
170. Schaumann, V., Influence of atropine and morphine on the liberation of acetylcholine from the guinea pig's intestine, *Nature (London)*, 178, 1121, 1956.

171. Paton, W. D. M. and Zar, M. A., The origin of acetylcholine released from guinea-pig intestine and longitudinal muscle strips, *J. Physiol. (London)*, 194, 13, 1968.
172. Kosterlitz, H. W., Opiate actions in guinea pig ileum and mouse vas deferens, in *Opiate Receptor Mechanisms*, Snyder, S. H. and Matthysse, S., Eds., MIT Press, Massachusetts, 1975, 68.
173. Creese, I. and Snyder, S. H., Receptor binding and pharmacological activity in the guinea pig intestine, *J. Pharm. Exp. Ther.*, 194, 205, 1975.
174. Snyder, S. H., and Pert, C. B., Membrane receptor, in *Opiate Receptor Mechanisms*, Snyder, S. H. and Matthysse, S., Eds., MIT Press, Massachusetts, 1975, 27.
175. Rónai, A. Z., Berzétei, I., and Bajusz, S., Differentiation between opioid peptides by naltrexone, *Eur. J. Pharmacol.*, 45, 393, 1977.
176. Shaw, J. S. and Turnbull, M. J., In vitro profile of some opioid pentapeptide analogues, *Eur. J. Pharmacol.*, 49, 313, 1978.
177. Leslie, F. M., Chvkin, C., and Cox, B. M., Ligand specificity of opioid binding sites in brain and peripheral tissues, in *Endogenous and Exogenous Opiate Agonists and Antagonists*, Way, E. L., Ed., Pergamon Press, New York, 1979, 109.
178. Goldstein, A., Tachibana, S., Lowney, L. I., Hunkapiller, M., and Hood, L., Dynorphin-(1-13), and extraordinarily potent opioid peptide, *Proc. Natl. Acad. Sci. USA*, 76, 6666, 1979.
179. Haycock, V. K. and Rees, J. M. H., The effect of morphine pretreatment on the sensitivity of mouse and guinea pig ileum to acetylcholine and morphine, *J. Pharm. Pharmacol.*, 24, 47, 1972.
180. Goldstein, A. and Schulz, R., Morphine tolerant longitudinal muscle strip from guinea pig ileum, *Br. J. Pharmacol.*, 48, 644, 1973.
181. Ehrenpreis, S., Greenberg, J., and Comaty, J. E., Mechanism of development of tolerance to injected morphine by guinea pig ileum, *Life Sci.*, 17, 49, 1975.
182. Ward, A. and Takemori, A. E., Studies on the narcotic receptor in the guinea pig ileum, *J. Pharm. Exp. Ther.*, 199, 117, 1976.
183. Cox, B. M. and Padhya, R., Opiate binding and effect in ileum preparations from normal and morphine pretreated guinea pigs, *Br. J. Pharmacol.*, 61, 271, 1977.
184. Johnson, S. M., Westfall, D. P., Howard, S. A., and Fleming, W. W., Sensitivities, of the isolated ileal longitudinal smooth muscle-myenteric plexus and hypogastric nerve-vas deferens of the guinea pig after chronic morphine pellet implantation, *J. Pharm. Exp. Ther.*, 204, 54, 1978.
185. Shoham-Moshonov, S. S. and Weinstock, M., Nature of seasonal variation in development of acute tolerance to morphine, *Eur. J. Pharmacol.*, 43, 153, 1977.
186. Schulz, R. and Herz, A., Aspects of opiate dependence in the myenteric plexus of the guinea pig, *Life Sci.*, 19, 1117, 1976.
187. Robson, L. E., Gillan, M. G. C., Waterfield, A. A., and Kosterlitz, H. W., The inhibitory effects of presynaptic adrenoceptor agonists on the contractions of the guinea pig ileum and mouse vas deferens in the morphine-dependent and withdrawn states, in *Characteristics and Function of Opioids*, van Ree, J. M. and Terenius, L., Eds., Elsevier/North-Holland Biomedical Press, Amsterdam, 1978, 67.
188. Gintzler, A. R., Serotonin participation in gut withdrawal from opiates, *J. Pharm. Exp. Ther.*, 211, 7, 1979.
189. Gintzler, A. R., Substance P involvement in the expression of gut dependence on opiates, *Brain Res.*, 182, 224, 1980.
190. Sjöstrand, N. O., The adrenergic innervation of the vas deferens and the accessory male genital glands, *Acta Physiol. Scand.*, 65, Suppl. 257, 1965.
191. Furness, J. B. and Iwayama, T., Terminal axons ensheated in smooth muscle cells of the vas deferens, *Z. Zellforsch.*, 113, 259, 1971.
192. Ambache, N. and Zar. A. M., Evidence against adrenergic motor transmission in the guinea pig vas deferens, *J. Physiol. (London)*, 216, 359, 1971.
193. Furness, J. B., Transmission to the longitudinal muscle of the guinea pig vas deferens: the effect of pretreatment with guanethidine, *Br. J. Pharmacol.*, 50, 63, 1974.
194. Henderson, G. and North, R. A., Depression by morphine of excitatory junction potentials in the vas deferens of the mouse, *Br. J. Pharmacol.*, 57, 341, 1976.
195. Marshall, I., Nasmyth, P. A., Nicholl, C. G., and Shepperson, N. B., α-Adrenoceptors in the mouse vas deferens and their effects on its response to electrical stimulation, *Br. J. Pharmacol.*, 62, 147, 1978.
196. Gibson, A. and Samini, M., The effects of bromocriptine on pre-synaptic and post-synaptic α-adrenoceptors in the mouse vas deferens, *J. Pharm. Pharmacol.*, 31, 826, 1979.
197. Illés, P., Zieglgänsberger, W., and Herz, A., Calcium reverses the inhibitory action of morphine on neuroeffector transmission in the mouse vas deferens, *Brain Res.*, 191, 511, 1980.
198. Waterfield, A. A., Lord, J. A. H., Hughes, J., and Kosterlitz, H. W., Differences in the inhibitory effects of normorphine and opioid peptides on the responses of the vasa deferentia of two strain of mice, *Eur. J. Pharmacol.*, 47, 249, 1978.

199. Szerb, J. C. and Vohra, M. M., Potencies of normorphine and Met-enkephalin in the vas deferens of different strains of mice, *Life Sci.,* 24, 1983, 1979.
200. Berti, F., Bruno, F., Omini, C., and Racagni, G., Genotype dependent response of morphine and methionine-enkephalin on electrically induced contractions of the mouse vas deferens, *Naunyn-Schmiedeb, Arch. Pharmacol.,* 305, 5, 1978.
201. Kosterlitz, H. W., Lord, J. A. H., Paterson, S. J., and Waterfield, A. A., Effects of changes in the structure of enkephalins and of narcotic analgesic drugs on their interactions with μ- and δ-receptors, *Br. J. Pharmacol.,* 68, 333, 1980.
202. Waterfield, A. A., Smockum, R. W. J., Hughes, J., Kosterlitz, H. W., and Henderson, G., In vitro pharmacology of the opioid peptides, enkephalins and endorphins, *Eur. J. Pharmacol.,* 43, 107, 1977.
203. Waterfield, A. A., Leslie, F. M., Lord, J. A. H., Ling, N., and Kosterlitz, H. W., Opioid activities of fragments of β-endorphin and of its leucine^{65}-analogue. Comparison of the binding properties of methionine- and leucine-enkephalin, *Eur. J. Pharmacol.,* 58, 11, 1979.
204. Hart, S. L., Kitchen, I., and Waddell, P. R., Different effects of current strength on inhibitory responses of the mouse vas deferens to methionine- and leucine-enkephalin, *Br. J. Pharmacol.,* 66, 361, 1979.
205. Wüster, M., Schulz, R., and Herz, A., Highly specific opiate receptors for dynorphin-(1-13) in the mouse vas deferens, *Eur. J. Pharmacol.,* 62, 235, 1980.
206. Schulz, R., Wüster, M., Krenss, H., and Herz, A., Selective development of tolerance without dependence in multiple opiate receptors of mouse vas deferens, *Nature (London),* 285, 242, 1980.
207. Illés, P., Zieglgänsberger, W., and Herz, A., Lack of cros-tolerance between morphine and Leu-enkephalin in the mouse vas deferens, *Brain Res.,* 177, 260, 1980.
208. Wüster, M., Schulz, R., and Herz, A., The direction of opioid agonists towards μ-, δ- and ϵ-receptors in the vas deferens of the mouse and the rat, *Life Sci.,* 27, 163, 1980.
209. Holmann, M. E., Nerve-muscle preparations of vas deferens, in *Methods in Pharmacology,* Vol. 3, Daniel, E. E. and Paton, D. M., Eds., Plenum Press, New York, 1975, 403.
210. Drew, G. M., Pharmacological characterization of the presynaptic α-adrenoceptor in the rat vas deferens, *Eur. J. Pharmacol.,* 42, 123, 1977.
211. Lemaire, S., Berube, A., Derome, G., Lemaire, I., Magnan, J., and Regoli, D., Synthesis and biological activity of β-endorphin and analogues. Additional evidence for multiple opiate receptors, *J. Med. Chem.,* 21, 1232, 1978.
212. Wüster, M., Schulz, R., and Herz, A., Specificity of opioids towards the μ-, δ- and ϵ-opiate receptors, *Neurosci. Lett.,* 15, 193, 1979.
213. Galoyan, A. A., Akopyan, T. N., Karapetyan, R. O., Arutunyan, A. A., and Organissyan, A. I., Enzymatic mechanisms of the formation of biologically active peptides in the hypothalamus, in *Endorphins '78,* Gráf, L., Palkovits, M., and Rónai, A. Z., Eds., Publishing House of the Hungarian Academy of Science, Budapest, 1978, 37.
214. Rónai, A. Z., Berzétei, I., Székely, J. I., Miglécz, E., and Bajusz, S., The in vitro pharmacology of D-Met^2, Pro^5-enkephalinamide, *J. Pharm. Pharmac.* submitted.
215. Wüster, M., Schulz, R., and Herz, A., Opioid agonists and antagonists: action on multiple opiate receptors, in *Endogenous and Exogenous Opiate Agonists and Antagonists,* Way, E. L., Ed., Pergamon Press, New York, 1979, 75.
216. Rónai, A. Z., Berzétei, I., and Kurgyis, J., Morphine lost the efficacy at but not the affinity to the opioid receptors of rat vas deferens, *Eur. J. Pharmacol.,* submitted.
217. Trendelenburg, U. and Haeusler, G., Nerve-muscle preparations of the nictitating membrane, in *Methods in Pharmacology,* Vol. 3, Daniel, E. E. and Paton, D. M., Plenum Press, New York, 1975, 457.
218. Bevan, J. A., Some functional consequences of variation in adrenergic synaptic cleft width and in nerve density and distribution, *Fed. Proc. Fed. Am. Soc. Exp. Biol.,* 36, 2439, 1977.
219. Enero, M. A., Langer, S. Z., Rothlin, R. P., and Stefano, F. J. E., Role of the α-adrenoceptor in regulating noradrenaline overflow by nerve stimulation, *Br. J. Pharmacol.,* 44, 672, 1972.
220. Knoll, J., and Illés, P., The isolated nictitating membrane of the cat. A new model for the study of narcotic analgesics, *Pharmacology,* 17, 215, 1978.
221. Hieble, J. P., Steinsland, O. S., Nelson, S. H., and Undesser, E. K., A new experimental model for studies on dopaminergic receptor mechanism, *Pharmacologist,* 18, 131, 1976.
222. Steinsland, O. S. and Hieble, J. P., Dopaminergic inhibition of adrenergic neurotransmission as a model for studies on dopamine receptor mechanisms, *Science,* 199, 443, 1978.
223. Steinsland, O. S., Furchgott, R. F., and Kirpekar, S. M., Inhibition of adrenergic neurotransmission by parasympathomimetics in the rabbit ear artery, *J. Pharm. Exp. Ther.,* 184, 346, 1973.
224. Hadházy, P., Vizi, E. S., Magyar, K., and Knoll, J., Inhibition of adrenergic neurotransmission by prostaglandin E$_1$ (PGE$_1$) in the rabbit ear artery, *Neuropharmacology,* 15, 245, 1976.

225. De la Lande, I. S. and Rand, M. J., A simple isolated nerve-blood vessel preparation, *Aust. J. Exp. Biol. Med. Sci.*, 43, 639, 1965.
226. Lazarus, L. H., Ling, N., and Guillemin, R., β-Lipotropin as a prohormone for the morphinomimetic peptides endorphins and enkephalins, *Proc. Natl. Acad. Sci. USA*, 73, 2156, 1976.
227. Bradbury, A. F., Smyth, D. G., and Snell, C. R., Liotropin: precursor to two biologically active peptides, *Biochem. Biophys. Res. Commun.*, 69, 950, 1976.
228. Gráf, L., Rónai, A.Z., Bajusz, S., Cseh, G., and Székely, J. I., Opioid agonist activity of β-lipotropin fragments: a possible biological source of morphine-like substances in the pituitary, *FEBS Lett.*, 64, 181, 1977.
229. Lewis, R. V., Stern, A. S., Kimura, S., Rossier, J., Stein, S., and Udenfriend, S., An about 50,000-dalton protein in adrenal medulla: a common precursor of /Met/- and /Leu/-enkephalin, *Science*, 208, 1459, 1980.
230. Li, C. H., Barnafi, L., Chretien, M., and Chung, D., Isolation and amino-acid sequence of β-LPH from sheep pituitary glands, *Nature (London)*, 208, 1093, 1965.
231. Hughes, J., Smith, T. W., Kosterlitz, H. W., Fothergill, L. A., Morgan, B. A., and Morris, H. R., Identification of two related pentapeptides from the brain with potent opiate agonist activity, *Nature (London)*, 258, 577, 1975.
232. Guillemin, R., Ling, N., and Burgus, R., Endorphines, peptides d'origine hypothalamique et neurohypophysaire a activite morphinomimetique, *C. R. Acad. Sci. Paris*, 282, 783, 1976.
233. Li, C. H. and Chung, D., Isolation and structure of an untriakontapeptide with opiate activity from camel pituitary glands, *Proc. Natl. Acad. Sci. USA*, 73, 1145, 1976.
234. Austen, B. M., Smyth, D. G., and Snell, C. R., γ-endorphin, α-endorphin and Met-enkephalin are formed extracellularly from lipotropin C-fragment, *Nature (London)*, 269, 619, 1977.
235. Aono, J., Takahashi, M., and Koida, M., An assay method of brain enzyme carving β-endorphin into methionine enkephalin, *Jpn. J. Pharmacol.*, 28, 930, 1978.
236. Gráf, L., Kenessey, Á., Berzétei, I., and Rónai, A. Z., Demonstration of β-lipotropin activating enzyme in porcine pituitary, *Biochem. Biophys. Res. Commun.*, 78, 1114, 1977.
237. Gráf, L., and Kenessey, Á., Specific cleavage of a single peptide bond (residues 77-78) in β-lipotrophin by a pituitary endopeptidase, *FEBS Lett.*, 69, 255, 1976.
238. Rossier, J., Bayon, A., Vargo, T. M., Ling, N., Guillemin, R., and Bloom, F., Radioimmunoassay of brain peptides: evaluation of methodology for the assay of β-endorphin and enkephalin, *Life Sci.*, 21, 847, 1977.
239. Liotta, A. S., Suda, T., and Krieger, D. T., β-endorphin is the major opioid-like peptide of human pituitary and rat pars distalis: lack of significant β-endorphin, *Proc. Natl. Acad. Sci. USA*, 75, 2950, 1978.
240. Gráf, L., Kenessey, Á., and Makara, G. B., Endorphins and/or artefacts: characterization of some pituitary proteinases involved in the generation of opioid peptides from β-lipotropin, in *Endorphins'78*, Gráf, L., Palkovits, M., and Rónai, A. Z., Eds., Publishing House of the Hungarian Academy of Science, Budapest, 1978, 127.
241. Gráf, L., Palkovits, M., and Rónai, A. Z., Eds., Tissue collection and extraction, in *Endorphins '78*, Publishing House of the Hungarian Academy of Sciences, Budapest, 1978, 197.
242. Gráf, L., Kenessey, Á., Patthy, A., Grynbaum, A., Marks, N., and Lajtha, A., Cathepsin D generates γ-endorphin from β-endorphin, *Arch. Biochem. Biophys.*, 193, 101, 1979.
243. De Wied, D., Kovács, G. L., Bohus, B., van Ree, J. M., and Greven, H. M., Neuroleptic activity of the neuropeptide β-LPH$_{62-77}$/(Des-Tyr¹/γ-endorphin; DTγE), *Eur. J. Pharmacol.*, 49, 427, 1978.
244. Rossier, J., Vargo, T. M., Minick, S., Ling, N., Bloom, F. E., and Guillemin, R., Regional dissociation of β-endorphin and enkephalin contents in rat brain and pituitary, *Proc. Natl. Acad. Sci. USA*, 74, 5162, 1977.
245. Watson, S. J., Akil, H., Richard III. C. W., and Barchas, J. D., Evidence for two separate opiate peptide neuronal systems, *Nature (London)*, 275, 226, 1978.
246. Ogawa, N., Panerai, A. E., Lee, S., Forsbach, G., Havlicek, V., and Friesen, H. G., β-endorphin concentration in the brain of intact and hypophysectomized rats, *Life Sci.*, 25, 317, 1979.
247. Cheung, A. L., and Goldstein, A., Failure of hypophysectomy to alter brain content of opioid peptides (endorphins), *Life Sci.*, 19, 1005, 1976.
248. Kangawa, K., Matsuo, H., and Igarashi, M., α-neoendorphin: a "big" Leu-enkephalin with potent opiate activity from porcine hypothalami, *Biochem. Biophys. Res. Commun.*, 86, 153, 1979.
249. Goldstein, A., Tachibana, S., Lowney, L. I., Hunkapiller, M., and Hood, L., Dynorphin-(1-13), an extraordinarily potent opioid peptide, *Proc. Natl. Acad. Sci. USA*, 76, 6666, 1979.
250. Mains, R. E., Eipper, B. A., and Ling, N., Common precursor to corticotropins and endorphins, *Proc. Natl. Acad. Sci. USA*, 74, 3014, 1977.
251. Mains, R. E., and Eipper, B. A., Studies on the common precursor to ACTH and endorphin, in *Endorphins '78*, Gráf, L., Palkovits, M., and Rónai, A. Z., Eds., Publishing House of the Hungarian Academy of Science, Budapest, 1978, 79.

252. **Nakanishi, S., Inoue, A., Taii, S., and Numa, S.,** Cell-free translation product containing corticotropin and β-endorphin encoded by messenger RNA from anterior lobe and intermediate lobe of bovine pituitary, *FEBS Lett.,* 84, 105, 1977.
253. **Gossard, F., Seidah, N. G., Crine, P., Routhier, R., and Chretien, M.,** Partial N-terminal amino acid sequence of pro-opio-melanocortin (ACTH/beta-LPH precursor) from rat pars intermedia, *Biochem. Biophys. Res. Commun.,* 92, 1042, 1980.
254. **Nakanishi, S., Inoue, A., Kita, T., Nakamura, M., Chang, A. C. Y., Cohen, S. N., and Numa, S.,** Nucleotide sequence of cloned cDNA for bovine corticotropin-β-lipotropin precursor, *Nature (London),* 278, 423, 1979.
255. **Nakanishi, S., Teranishi, Y., Noda, M., Notake, M., Watanabe, Y., Kakidani, H., Jingami, H., and Numa, S.,** The protein-coding sequence of the bovine ACTH-β-LPH precursor gene is split near the signal peptide region, *Nature (London),* 287, 752, 1980.
256. **Voigt, K. H. and Weber, E.,** Corticotropin, β-lipotropin/β-endorphin and the non-corticotropin/β-lipotropin portion of their common precursor are present in the same secretory granules, in *Endogenous and Exogenous Opiate Agonists and Antagonists,* Way, E. L., Ed., Pergamon Press, New York, 1979, 313.
257. **Rossier, J., Audigier, Y., Ling, N., Cros, J., and Undenfriend, S.,** Met-enkephalin-Arg[6]-Phe[7], present in high amounts in brain of rat, cattle and man, is an opioid agonist, *Nature (London),* 288, 88, 1980.

Chapter 2

THE MOST CHARACTERISTIC IN VIVO EFFECTS OF OPIATES

J. I. Székely

TABLE OF CONTENTS

I. Introduction 30

II. Antagonism by Naloxone 30
    A. Is the Observed Antagonism of Competitive Nature or Not? 31
    B. Are the Opiate Receptors Located in Different Organs or Different Parts of the Brain Homogeneous? 33
    C. Do the Different Narcotic Drugs Interact with the Same Type of Opiate Receptors? 35
    D. Does the Affinity of Opiate Receptors for Agonists and Antagonists Change Under Physiological Conditions and Upon Pharmacological Manipulation? 37

III. Stimulus Properties 38
    A. Stimulus Properties in Humans 38
    B. Stimulus Properties in Experimental Animals 39
    C. Is It Possible to Suppress the Stimulus Properties of Narcotic Drugs by Modifying the Nonopiate Functions of the Brain? 41

IV. Analgesic Effect 42

V. Behavioral Effects 45
    A. Changes in the Spontaneous Motor Activity 46
    B. Catalepsy, Rigidity 51
    C. Stereotypy 52

VI. Drug Interaction Studies 53
    A. Drugs Affecting Brain Catecholamines and Serotonin 53
    B. Cholinolytics and Cholinomimetics 63
    C. Other Drugs 65

VII. Localized Intracerebral Application 66

VII. Effects of Opiates on Conditioned Behavior 73

IX. Modulation of the Turnover of Neurotransmitters and Some Other Substances 75
    A. Effects on Acetylcholine (ACH) Turnover 75
    B. Effect on Serotonin (5-HT) Turnover 80
    C. Effect on Catecholamine Turnover 84
    D. Effects on Dopamine Sensitive Cyclic Nucleotides 88
    E. Effect on Adenosine Triphosphatase (ATPase) Activity 91
    F. Effect on the Brain Calcium and Lipids 92
    G. Other Metabolic Effects 93

References 95

I. INTRODUCTION

Considering the multiplicity of receptors and diversity of other biochemical mechanisms, which may participate in mediation of the pharmacological effects elicited by different drugs, the question arises: is it possible at all to study the opiate receptor functions in vivo? How can we find out that the "characteristic" symptoms observed after administration of opiates and opioid peptides are really opiate receptor mediated effects? To expect that all the biological responses observed upon administration of opiates are mediated by opiate receptors would surely be an oversimplification. If this proposition were true all the opioids should elicit qualitatively similar effects, which is surely not the case. Moreover, opiates have many sites of action, i.e., they may modify the synaptic transmission in very different parts of the nervous system. Thus the direct effects, i.e., biological actions which are primary consequences of opiate receptor stimulation are frequently confounded by the indirect effects. Indirect are regarded those actions which are secondary manifestations of events specifically mediated by opiate receptors. Furthermore, both the direct and indirect effects may disturb the normal homeostasis of the living organism, which in turn initiates a series of adaptive mechanisms. Some of them are apparent while the opiate is still in the organism while others persist for longer periods and apparently do not coincide with the receptor occupation. Sometimes it is extremely difficult to differentiate between the secondary outcome of opiate specific effects and the nonspecific adaptive mechanisms. The supersensitivity of nonopiate receptors is probably an adequate example for the formers and the stressor effects of opiates for the latters. Events become even more difficult to analyze if morphine-like drugs are administered chronically.

Reduced food intake (weight loss in chronic experiments) may be a manifestation of the particular subjective state elicited by opiates characterized among others by a reduction in motivation. However, it also may be a consequence of decreased pancreatic secretion brought about directly by local opiate receptors. Moreover it can be attributed to the activation of the hypothalamic satiety center upon alteration of acetylcholine or catecholamine release by opiate receptors. Or it may be a secondary result of decreased intestinal motility and secretion brought about by peripheral and central modes of action. In the case of chronic administration it also may be symptom of the daily withdrawal syndrome preceding the next injection of narcotic, moreover a consequence of adrenal exhaustion due to the repeated stressor effect of multiple opiate injections.

Nevertheless, there are some specific actions observable in behavioral experiments or in human studies, which may reliably be used to establish whether the observed pharmacological effects have been mediated by opiate receptors or not.

In this chapter attention will be focused to the opiate specific behavioral and metabolic effects in vivo. Some of their autonomic and electrophysiological effects are also very characteristic but in order to save space and time they will be discussed together with the more or less similar actions of opioid peptides.

II. ANTAGONISM BY NALOXONE

Just as in the opiate sensitive isolated organ preparations, antagonizability by naloxone of any behavioral effect proves its specificity. Reversal by narcotic antagonists was considered an *a priori* criterium of opiate-specific actions long before the discovery of opiate receptors. As postulated by Arunlakshana and Schild[1] in 1959 " . . . Although drug receptors have not so far been identified by physical or chemical methods, they can be identified pharmacologically by means of antagonists. If two agonists act on the same receptors they can be expected to be antagonized by the same concentration of antagonist and to produce the same pA_x . . . pA_x values can be used to identify agonists, which act on the same receptors".

In carefully designed behavioral experiments the reversal of opiate specific effects may be demonstrated almost as exactly as in vitro. Moreover even the so called "apparent pA_2" values can be calculated just as by means of the classical isolated organ preparations in vitro. The validity and importance of the above-cited statements of Arunlakshana and Schild[1] has become really obvious only by the discovery of opiate and other receptors. Until that time the receptor theory and receptor kinetics were only mathematical models, which made possible to describe the data of pharmacological experiments with precision. Now, as the previously only theoretically existing receptors have been materialized, the data recorded under in vitro conditions may be interpreted in more or less concrete chemical and physical terms. Of course the data obtained in vivo should be interpreted even in the future with caution as differences in metabolic stability, absorption capacity and ability to pass the capillary wall and blood brain barrier moreover differential distribution in the different tissues may distort the in vivo recorded (quantitative) data.

Surveying the relatively scanty literature on in vivo determination of pA_2 values it may be concluded that the significance of pharmacodynamic studies in living animals has not been appreciated appropriately. Although by determining the pA_2 values in vivo some very important questions can be clarified in the living organism under natural conditions, i.e., without killing the animals and without destruction of normal tissue and cell structure by dissection, chopping, homogenization. The most important problems to be discussed are as follows.

A. Is the Observed Antagonism of Competitive Nature or Not?

In the former case a dose-dependent parallel shift of the dose-response curve to the right can be demonstrated in vivo just as in the classical in vitro studies.

As Figure 1 shows, in mouse hot plate test morphine elicits a dose-related lengthening of paw-licking latency, i.e., it increases the pain threshold. Simultaneous treatment with naloxone results in a parallel shift of the dose-response curve obtained by transforming the original numerical values into nonparametric form. Thus the criterion of parallel shift of dose-response curve of agonist by the antagonist is fulfilled. Consequently the antagonism is of competitive type. Similar competitive antagonism can be demonstrated under the same experimental conditions using a potent enkephalin analog (D-Met, Pro-EA) or β-EP as agonists and naloxone as antagonist (Figures 2, 3). Thus as for the analgesic action of morphine, D-Met, Pro-EA, and β-EP, it may be regarded an opiate receptor mediated action, since the main criterium of opiate specific effects is their antagonizability by naloxone in competitive manner.

An example of the pharmacological effects induced by narcotic agonists but not mediated by opiate receptors may be the meperidine elicited hypothermia, which contrary to the similar action of methadone can not be reversed by nalorphine.[2]

Otherwise, the majority of the behavioral, electrophysiological, metabolic, and autonomic effects elicited by narcotics can be antagonized by naloxone-like drugs competitively. It has been proved in the case of morphine-, methadone-, levorphanol-, meperidine-induced analgesia in mouse hot plate test[2] and in writing assay using morphine, etorphine, levorphanol, pentazocine, cyclazocine, nalorphine, or methadone as agonists and naloxone or diprenorphine as antagonists[3,4] and similarly in rat tail-flick test.[5]

Of the autonomic effects induced by narcotics lenticular opacity,[2,6] inhibition of gastrointestinal motility,[2] respiratory depression[6,7] have been found antagonizable by

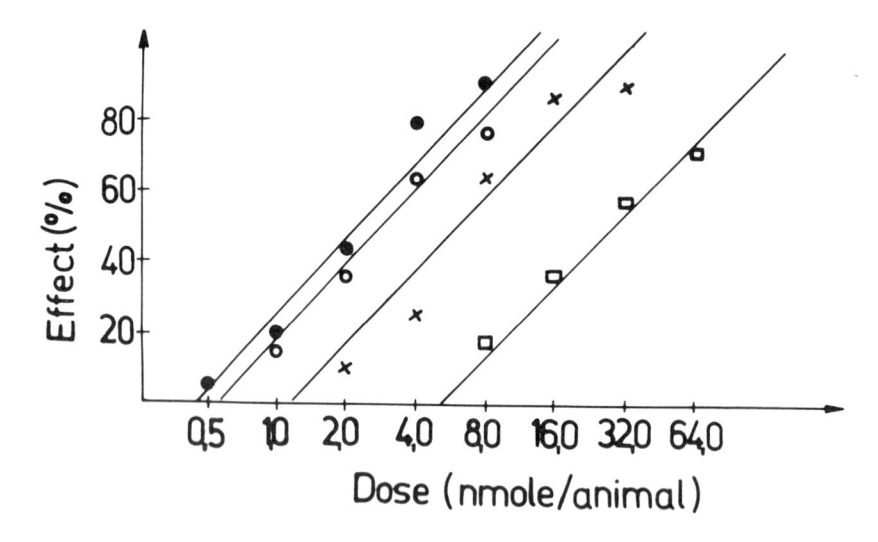

FIGURE 1. Effects of 0.01 (O), 0.10 (X), and 1.0 (□) mg/kg naloxone pretreatment (s.c.) on the analgesic action of i.c.v. administered morphine (●) in the mouse hot plate test. Dose response curves. Animals were considered positive for analgesia if the control latency time was at least doubled upon treatment. Effect means per cent of animals positive for analgesia. At each dose level at least 15 mice were used.

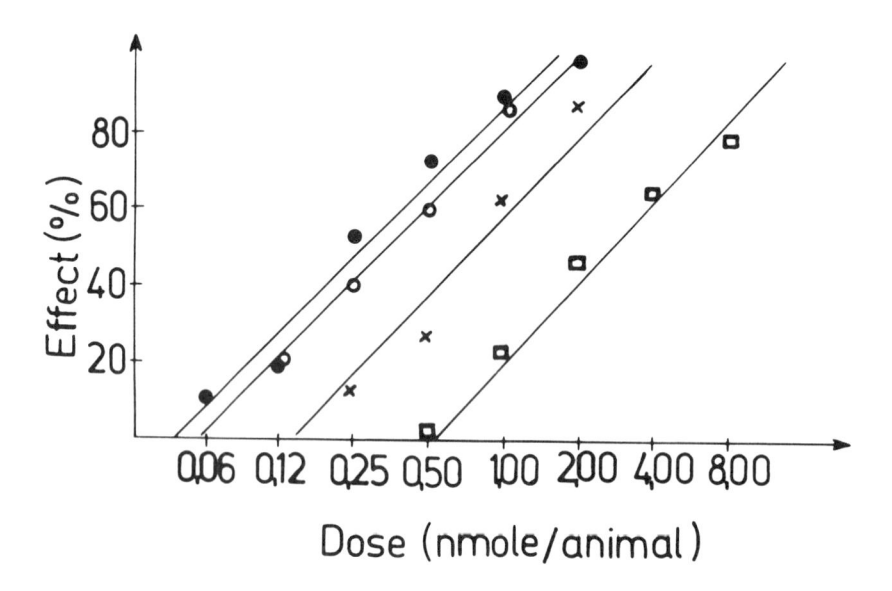

FIGURE 2. Effect of naloxone pretreatment on the analgesic action of β-EP. See Figure 1 for further details.

naloxone. (The antagonism of electrographic effects induced by opiates will be discussed later.) Nevertheless, very many well-known actions of opiates regarded generally specific for this group of drugs have not been examined yet as for their antagonizability by narcotic antagonists. This shortage in information is probably not due to the lack of interest on the part of the pharmacologists, rather, it may be attributed to the time-consuming nature of such kind of experimentation. Nevertheless, determination of the pA_2 value is not an absolute requirement in testing the contingent competi-

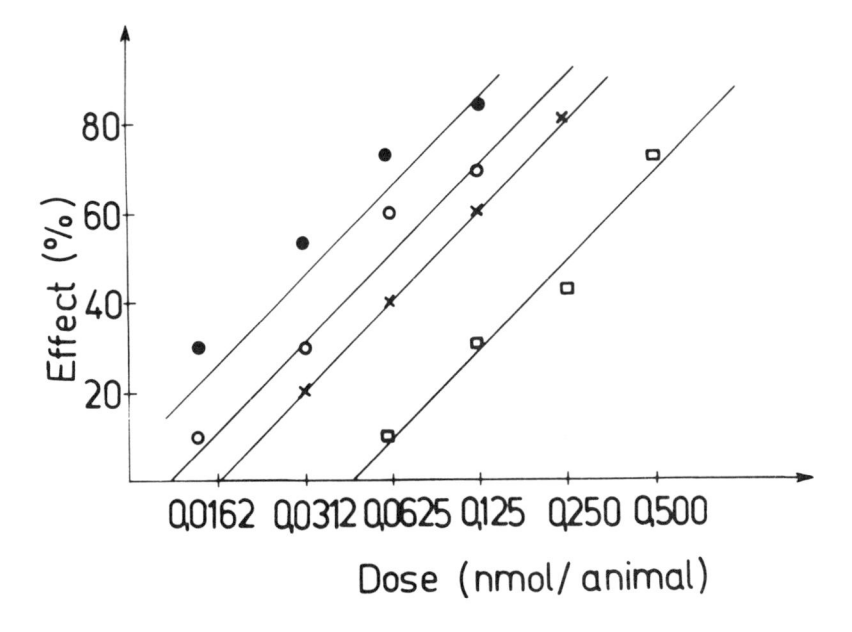

FIGURE 3. Effect of naloxone pretreatment on the analgesic action of D-Met, Pro-EA. See Figure 1 for further details.

tive nature of interaction. From this point of view the presence or absence of parallel shift of the dose-response curves is in itself sufficient.[1] However, the determination of the apparent pA_2 values can give further valuable information.

B. Are the Opiate Receptors Located in Different Organs or Different Parts of the Brain Homogeneous?

If the different pharmacological effects of an agonist are mediated by the same type of receptors, similar pA_2 values are expected when using the same antagonist. The pA_2 value according to its definition is the negative logarithm of molar dose of antagonist, which reduces the effect of a double dose of agonist to that of a single dose. (High pA_2 means that only a small amount of antagonist is needed to diminish the potency of agonist by 50%, while low pA_2 value shows that much of it must be administered to attain the same effect.) If in two different test systems similar pA_2 values are obtained upon using the same pair of agonist and antagonist, the receptors under study are of the same type.

Let us suppose that the opiate receptors in the brain and the gut are similar, moreover that the access of drugs to the receptors is the same in the above-mentioned organs and also their accumulation and elimination are similar. In this case similar pA_2 values are expected just as with two in vitro preparations containing the same receptors. Thus determination of the apparent pA_2 values is a powerful tool in examining whether the opiate receptors located in different parts of the nervous sytem are similar or not.

Table 1 summarizes the pertinent data available in the literature in relation of pA_2 values of some narcotic antagonists against morphine in different in vivo experimental paradigms and in different species.

Probably the most striking outcome of this comparison is the similarlity in the pA_2 values of naloxone against morphine in different analgesic assays such as tail-flick, hot plate, writhing, and shock titration (see V, VII, IX-XII, XIV-XV, XVIII-XX, XXIII-XXIV in Table 1). Thus it seems to more or less proven that the different con-

Table 1
COMPARISON OF THE pA$_2$ VALUES OF NARCOTIC ANTAGONISTS AGAINST MORPHINE[a] IN DIFFERENT IN VIVO PHARMACOLOGICAL ASSAYS

Serial number	Antagonist	Species	Assay	pA$_2$	Ref.
I	Nalorphine s.c.	Mouse	Hot plate	6.18	2
II	Nalorphine s.c.	Mouse	Induction of lenticular opacity	5.92	2
III	Nalorphine s.c.	Mouse	Writhing	6.71	3
IV	Nalorphine s.c.	Mouse	Inhibition of intestinal motility	5.63	3
V	Naloxone s.c.	Mouse	Writhing	7.01	3
VI	Naloxone s.c.	Mouse	Inhibition of intestinal motility	6.60	3
VII	Naloxone s.c.	Mouse	Writhing	7.07	4
VIII	Diprenorphine s.c.	Mouse	Writhing	7.73	4
IX	Naloxone s.c.	Mouse	Writhing	7.10	8
X	Naloxone s.c.	Mouse	Hot plate	7.27	8
XI	Naloxone s.c.	Mouse	Tail flick	7.05	8
XII	Naloxone s.c.	Mouse	Hot plate (55°C)	6.90	9
XIII	Nalorphine s.c.	Mouse	Hot plate (55°C)	5.80	9
XIV	Naloxone i.p.	Rat	Electrical stimulation of the tail	6.80	10
XV	Naloxone s.c.	Mouse	Tail flick	7.05	7
XVI	Naloxone s.c.	Mouse	Depression of respiration	7.35	7
XVII	Naloxone i.p.	Rat	Induction of hyperthermia	6.57	7
XVIII	Naloxone i.m.	Monkey	Shock titration	7.16	11
XIX	Naloxone i.p.	Rat	Hot plate	7.04	12
XX	Naloxone i.p.	Rat	Tail flick	6.98	12
XXI	Naloxone intrathecally	Rat	Hot plate	8.05	12
XXII	Naloxone intrathecally	Rat	Tail flick	8.20	12
XXIII	Naloxone s.c.	Mouse	Hot plate	6.86	5
XXIV	Naloxone s.c.	Rat	Tail flick	7.17	5

[a] Morphine given systemically (s.c., i.p., i.m.) but in experiments N° XIX, XX intrathecally.

ventionally applied tests used in measurement of antinociceptive action of narcotic drugs reflect the same receptor mechanism. A not very trivial finding if considering that the neural pathways subserving such different movements as tail withdrawal (tail flick test), licking of the paws (hot plate test), pressing down a lever (shock titration), stretching of the abdominal wall (writhing test) are obviously not the same.

A further unexpected finding is the similarity of apparent pA$_2$ values obtained in different species (mouse, rat, monkey) upon measurement of the antinociceptive action (Table 1). Thus the receptors mediating the analgesic effect of morphine should be similar in these species.

However, certain differences were found upon comparing the pA$_2$ values of naloxone obtained during testing for analgesia to those determined in relation of inhibition of intestinal motility (III vs. IV and V vs. VI in Table 1) or eliciting lenticular opacity (I vs. II) or in the case of respiratory depression (XV vs. XVI). In other words more naloxone is needed to antagonize the morphine-induced intestinal inhibition[3] and lenticular opacity[2] than to reverse the analgesic effect and less is sufficient to prevent the respiratory depressant effect.[7]

Very high pA$_2$ values were obtained upon intrathecal naloxone and systemic morphine administration (XXI, XXII) but relatively low ones (XIX, XX) if giving morphine intrathecally and naloxone i.p.,[12] proving the importance of route of administration.

Finally Table 1 demonstrates significant differences in pA$_2$ values if different antagonists are used with the same agonists (I vs. V, IV vs. VI, and XII vs. XIII). It is probably not surprising that of nalorphine relatively larger amount is needed to elicit the same degree of antagonism than of naloxone, i.e., the pA$_2$ values of the latter are greater against the same agonists. As nalorphine is a mixed agonist-antagonist while naloxone is one of the purest (and strongest) antagonists known to date,[13] the observed differences in potency correspond to that, what is theoretically expected.

Consequently the opiate receptors mediating the analgesic effect of morphine and those responsible for the autonomic effects are different. A fact which the pharmacologists were not very conscious of, but as seen above it has been sufficiently proven already before the discovery of endogenous opioids. Thus the multiplicity of opiate receptors may be considered as proven on the basis of data accumulated during the last two decades. Nevertheless introduction of new in vitro opiate sensitive techniques (Chapter 1, Volume I) and the detailed comparative analysis of opiates and opioid peptides were necessary to convince the research workers of the heterogeneity of opiate receptors.[14-20]

C. Do the Different Narcotic Drugs Interact with the Same Type of Opiate Receptors?

This question represents another approach of the same problem concerning the uniformity or multiplicity of opiate receptors. Even if acknowledging the participation of different opiate receptors in mediation of different opiate specific actions the question seems to be pertinent: are the same specific actions of different morphine congeners (e.g., analgesia, behavioral symptoms, respiratory depression, etc.) brought about by the same subgroups of receptors or not? It is conceivable that the relative affinities of various opioids to the distinct subclasses of opiate receptors are not the same resulting in quantitative (and qualitative) differences in their effects and the corresponding ED$_{50}$ values. However, this does not exclude the possibility that at least a part of the opiate specific actions are mediated by the same subgroup of receptors, e.g., analgesia. If it is true, in spite of the vastly different analgesic ED$_{50}$ values, similar pA$_2$ values will be expected using any opiate antagonist against different agonists.

As for the majority of classical morphine surrogates, this assumption seems to be valid. Smits and Takemori[21] have compared the pA$_2$ of naloxone against morphine, methadone, and levorphanol in the mouse phenylquinone test. The s.c. analgesic ED$_{50}$ values were found as 0.27, 0.40, and 0.059 mg/kg, respectively. The pA$_2$ values of naloxone against them were measured as 7.08, 6.87, and 6.98, respectively.[21] That is the analgesic potencies of the three narcotics were found significantly different, but their antinociceptive actions could be antagonized by practically the same amount of naloxone. In other words the same receptor population (mechanism) mediates the analgesia induced by the selected natural (morphine), synthetic (methadone), and semisynthetic (levorphanol) narcotics. Up to the present no data have been published which would contradict this conclusion at least in relation of the classical alkaloids and their synthetic derivatives. However, two exceptions must be put forward in advance.

The pA$_2$ values of naloxone against the above-mentioned "pure" agonists were found to be different from those against some mixed agonist-antagonists such as pentazocine, cyclazocine, and nalorphine.[4,21] In the latter cases the pA$_2$ values were 6.20, 6.50, and 6.21, i.e., significantly lower. Calculating the corresponding antilogarithms it turns out that about 5 to 7 times more naloxone is needed to antagonize the analgesic effect of mixed agonist-antagonist than that of the pure agonists as shown by Smits and Takemori.[21]

These differences in pA$_x$ plots for the two groups of narcotics can be explained in different ways.

The same authors reported[4,21] that the slopes of the naloxone's pA$_x$ plots were greater than unity if using nalorphine, cyclazocine, or pentazocine as agonists. Thus as put forward by Arunlakshana and Schild,[1] either a failure of the drugs to attain equilibrium with the receptors or a paradoxical potentiating effect may be supposed.

The other possibility is to suppose the existence of a separate subgroup of receptors, where the nalorphine-like drugs act as agonists and naloxone as antagonist. Experimental data obtained by Martin et al.[22,23] in chronic spinal dogs may support such an explanation. Working exclusively in spinal dogs they analyzed in detail the interaction of different agonists and antagonists. They concluded on the existence of three different classes of opiate receptors[22,23] (μ, $\varkappa$, and σ). According to their hypothesis morphine would be a strong agonist of both μ- and $\varkappa$-receptors, nalorphine, cyclazocine, and pentazocine, however, competitive antagonists of the formers and agonists of the latters. According to their studies ketocyclazocine is a selective agonist of $\varkappa$-receptors while naloxone is a pure antagonist of both μ- and $\varkappa$-receptors, but about 20 times weaker on the latter ones. Thus it is not surprising that much more naloxone is needed to antagonize the nalorphine-induced analgesia than that of the morphine-like drugs.

Interestingly in a recent study Tortella et al.[24] found the electrographic effects of ethylketocyclazocine more sensitive to naloxone than those of morphine. This finding (obtained in rats) confirms the differential antagonizability by naloxone of the morphine and ethylketocyclazocine induced actions, but the direction of these differences is opposite to those recorded in spinal dogs.[23]

Martin and co-workers in their already classical experiments[22,23] did not determine the pA_2 values. Their theory on the receptor multiplicity was based mainly on their findings that the so-called $\varkappa$-receptor agonists (e.g., ketocyclazocine) did not suppress the abstinence syndrome in morphine-dependent spinal dogs. Thus the selective $\varkappa$ agonists have no significant affinity to the μ-receptors. Moreover, the ketocyclazocine-dependent dogs were not tolerant to morphine. Naloxone, naltrexone, or diprenorphine induced abstinence both in morphine and ketocyclazocine-dependent dogs but in the latter they were 20 to 60 times less potent in this aspect. The third subclass of receptors whose specific agonists were N-allylnorphenazocine had been designated σ-receptors.[22,23]

A nice hypothesis, which is getting more and more frequently quoted in the recent studies connected with the pharmacology of enkephalins and endorphins. But the original proposals of Martin and co-workers[22,23] have not been confirmed by direct measurement of the corresponding pA_2 values. Cowan et al.[25] found the pA_2 values of naloxone against morphine and ethylketocyclazocine similar in the tail compression test, which contradicts the hypothesis on the multiplicity of opiate receptors. It is conceivable that morphine and this benzomorphane derivative act on different receptors in dogs but on similar ones in rats; since species differences of the opiate receptors have been amply demonstrated in vitro and in vivo (Chapter 3, Volume III)

Recently Schulz et al.[20] postulated the existence of a further type of opiate receptors in rat vas deferens, which were specifically sensitive to β-EP, but almost insensitive to morphine, natural enkephalins, etc. These receptors are designated as ε. Unfortunately no in vivo data are available on these ε-receptors.[19,20] Just as in relation of the δ-receptor population detected first by Lord et al.[16] and Waterfield et al.[17] in the mouse vas deferens (see Chapter 1, Volume III).

A third possibility is that a uniform type of receptor interacts in different ways with the naloxone-and morphine-like drugs. This assumption surely seemed to be too hypothetical until the discovery of opiate receptors. However, considering that certain alterations of the ionic composition of extracellular space affect the receptor binding of agonists and antagonists in opposite directions (Chapter 1, Volume I), even this possibility must be taken into consideration. Thus, it is well established that increasing the concentration of Mg⁺⁺ or decreasing the Na⁺ level, the relative affinities of opioid agonists to their receptors are increased while in opposite cases the binding of antagonists is facilitated (Chapter 1, Volume I and Chapter 1, Volume II). This relationship proved to be valid for all opiates and opioid peptides studied so far, independently of their chemical structure (Chapter 1, Volume I and Chapter 1, Volume II). Thus, these anorganic ions affect the receptors themselves and not their ligands. These findings infer, however, that the opioid agonists and antagonists interact with the same receptors at least in part in different manner. Thus, there may be several different explanations of the anomalous behavior of mixed agonist-antagonists upon determining the pA$_2$ values of naloxone against them. Consequently the above-discussed hypothesis on uniformity of receptors mediating such a carefully selected opiate specific action as analgesia may not be regarded either unambiguously confirmed or refuted. Probably the detailed pharmacological analysis of opioid peptides will provide additional data in this aspect too. Unfortunately our experiments[5,18] make the matter only more complicated (Chapter 3, Volume II).

A further interesting question which emerged in connection with the pA$_2$ determinations is as follows.

D. Does the Affinity of Opiate Receptors for Agonists and Antagonists Change Under Physiological Conditions and Upon Pharmacological Manipulations?

By giving an appropriate answer to this question many yet unsolved problems of the pharmacology could be clarified, but until now little progress has made in this direction.

Harris et al.[26] examined the effect of i.c.v. administred Ca⁺⁺ on the analgesic effect of morphine and its reversal by naloxone. As expected, Ca⁺⁺ administration attenuated the antinociceptive effect of morphine in a dose-dependent way, but the pA$_2$ value of naloxone did not change.[26] Thus this alteration in the ionic composition of the brain decreased the efficacy of narcotic agonist without changing the potency of antagonist under study. On the other hand, single morphine treatment may change the sensitivity of opiate receptors to naloxone without altering its reactivity to agonists. In experiments reported by Takemori et al.[27] mice were pretreated with morphine and after 2 hr, when its analgesic effect had already dissipated, the apparent pA$_2$ of naloxone against morphine was (re)determined. In these mice, pretreated with morphine, the pA$_2$ value shifted from 6.96 to 7.30 though the analgesic ED$_{50}$ of morphine did not change. This alteration of the apparent pA$_2$ lasted for several days. Pretreatment with a mixed agonist-antagonist pentazocine did not elicit similar alteration of the pA$_2$ value.[27]

These findings were later confirmed and extended to other agonists and antagonists.[28] Pretreatment of mice with other analgesics such as levorphanol and methadone also increased the potency of naloxone.[28] On the other hand, morphine pretreatment increased the potency of other antagonists too, i.e., those of nalorphine and diprenorphine.[28] The same authors[29] supposed that this increment in the potency of naloxone was connected with tolerance development (to the action of agonists). It has been repeatedly demonstrated that upon tolerance development also the pA$_2$ values of antagonists increased. In opiate-tolerant rats the pA$_2$ value of naloxone was found significantly elevated.[29] Moreover, in striatal slices of mice, prepared 3 days after implantation of morphine pellet, naloxone induced in smaller concentration morphine release than from similar preparation of naive animals.[30] This finding implies that in opiate-tolerant animals the binding of agonists is selectively reduced. However, the above-mentioned enhanced sensitivity to naloxone following a single treatment without agonist cannot be regarded simply as an expression of tolerance development. The morphine-induced tolerance induction is a slower process, generally not detectable earlier than 1 day after morphine administration.[29] However, the single morphine treatment caused elevation of pA_2 value reaches its peak about 2 to 3 hr after its administration.[27,28] Consequently, the increase in the efficacy of naloxone upon single treatment of animals with opiate agonists may be observed still before tolerance development. Furthermore, the efficacy of the antagonist rises much faster than the development of tolerance.[29] Moreover, cycloheximide, a protein synthesis inhibitor known to inhibit both tolerance development (see Chapter 9, Volume III for literature) and the accompanying increment in the potency of naloxone,[29] did not reverse the enhancement of naloxone's potency in acute experiments.[31] Thus the two processes, namely tolerance development and increase in the pA_2 values, do not covary. Nevertheless, if trying to interpret the above-mentioned sensitization to naloxone upon morphine treatment, the possibility of a very simple explanation also must be taken into account. Several authors[32,33] reported on facilitated disposition of naloxone to the brain following acute or chronic morphine treatment. Obviously the pharmacological action of opiates are more easily inhibited if the antagonist enters into the brain in greater amount.

It may be concluded that pA_2 determination is a powerful tool in determining the specificity of certain effects of opiates, in classification of opiate receptors, and in characterization of the latter's functional state in vivo. The inherent possibilities of this method are far from being exhausted yet. (Unfortunately, the pA_2 determination in vivo is very tiring, which probably explains why this method so informative has been applied relatively rarely.)

Another method of identification of opiate specific actions is probably newer than the quantitative study of antagonism yet relatively widely used.

III. STIMULUS PROPERTIES

Morphine and related narcotics induce a characteristic syndrome of subjective effects which humans (especially narcotic addicts) and experimental animals can easily differentiate from the psychic effects induced by other psychoactive drugs.

A. Stimulus Properties in Humans

As discussed in the pharmacological textbooks, morphine and its congeners elicit characteristic subjective sensations also in humans. Many of these symptoms are related to their euphorisant actions such as a feeling of well-being and satisfaction, a definite anxiolytic effect, feeling of efficiency, enhanced self image, a feeling of complete harmony with the world.[34] Other subjective symptoms are connected with the sedative properties of these drugs and a third group of specific symptoms reflect their histamine liberating and autonomic effects such as nausea, dizziness, loss of appetite, etc. Probably due to the latter effects, the morphine-induced subjective sensations are not necessarily pleasant and opiates have reinforcing, addictive character only in a part of healthy humans.[34] However, independently of being addictive or nonaddictive for the different individuals, their subjective effects can easily be differentiated from those induced by other types of psychoactive drugs. Using questionnaires consisting of adequately compiled items relating to the subjective state induced by morphine-like drugs, it can be distinguished from those elicited by pentobarbitone, chlorpromazine and alcohol, or the nalorphine-cyclazocine group, or LSD.[35-37] Interestingly it seems to be fairly difficult to differentiate the subjective effects of morphine and amphetamine.[37] Thus, in some cases the subjective effects induced by morphine and amphetamine derivatives were summarily examined.[34]

Interpretation of these studies already lies outside the scope of this book. Nevertheless, some points must be raised which emerge upon synthesizing the data obtained.

Concerning some individual items of these questionnaires the morphine and amphetamine derivatives can easily be differentiated from others. However, examining the answers given to the questions relating to the two groups, these drugs cannot be differentiated.[34]

A further difficulty arises if the doses applied are also varied. The mixed opiate agonist-antagonists (nalorphine, cyclazocine, etc.) elicit symptoms characteristic of morphine in small doses, while at higher doses the subjective sensations elicited are more similar to those induced by psychomimetics.[34]

It may be concluded that opiate agonists induce characteristic subjective effects which may be reliably distinguished from those induced by other types of psychoactive drugs if adequately examined. The activation of opiate receptors may be objectively detected in vivo in humans by experimental methods. Thanks to the witty method introduced by Overton[38] the subjective effects of opiate receptor activation may be examined in experimental animals too.

B. Stimulus Properties in Experimental Animals

Overton[38] showed, in his already classical studies, that drug treatment in itself might serve as discriminative stimulus in behavioral experiments. Rats can be trained to choose the right or left arm of a T maze, or one of the two levers in Skinner box depending on the kind of treatment applied prior to the experimental session. In these experiments obviously the interoceptive sensations elicited by the drug treatment induced autonomic reactions and/or its direct central effects served as discriminative cue. This method turned out to be a very effective tool for in vivo recording of activation of the opiate receptors and in delineation of the narcotic drugs as well.

In experiments of Hill et al.[39] rats were trained to escape footshock in a T maze by turning one way under a drug and the other way upon saline treatment without any aid of exteroceptive cues. Giving saline and the different doses of morphine alternatively in balanced order, a highly significant, dose-related state-dependent behavior could be established.[39] In view of this interesting finding on the discriminative cuing ability of morphine, many questions arise and hitherto few of them have been extensively studied. These are as follows:

Can other narcotic drugs also serve as discriminative stimuli?

The simple answer is yes. All narcotic drugs may serve as discriminative stimuli if they reach the central receptors.[40] The lack of discriminative property in the case of loperamide is obviously due to its inability to penetrate beyond the intestinal wall.[41,42]

Does the stimulus property of one narcotic drug generalize to others?

This is the only extensively studied aspect of this phenomenon. Obviously the presence or absence of generalization of the stimulus property may be connected with the identity or diversity of receptors mediating their effects. Numerous data indicate that there are no differences in the discriminative stimulus properties of narcotic analgesics, i.e., they can substitute each other in the above-mentioned experimental paradigms.[40,43-45] According to these studies[40,43-45] morphine, codeine, heroin, phenazocine, oxymorphone, levorphanol, meperidine, methadone, fentanyl, etonitazen, bezitramide, and dextromoramide constitute a common group in spite of the vast differences in their chemical structure. These observations are in good agreement with the general clinical experience that in humans one opiate may be easily substituted with another narcotic agonist. These results suggest that the property enabling morphine to function as discriminative stimulus in the rat is analogous to the component of action of morphine responsible for producing characteristic subjective effects in humans. It is very interesting that ketocyclazocine failed to substitute for the training dose of morphine in rats.[44] As both are opiate agonists, it is tempting to suppose that the differences in their stimulus properties are due to the differences in the receptor populations activated by them. Thus this experiment seems to confirm the previously mentioned hypothesis on the multiplicity of opiate receptor mechanisms in vivo.[22,23]

However, morphine-like agonists cannot be substituted with naloxone, nalorphine, levallorphan, or oxylorphan, i.e., with structurally related narcotic antagonists.[44] Moreover, administering morphine with naloxone, the former loses its stimulus property, i.e., upon combined treatment the animals chose the "saline-correct" lever.[43] Thus, their stimulus property is connected with the activation of opiate receptors and not solely with receptor occupancy. As for the substitution of morphine or fentanyl with mixed agonist-antagonists, the data are conflicting. In a study of Colpaert et al.,[46] giving nalorphine or cyclazocine to rats trained to discriminate between fentanyl and saline, the fentanyl-correct lever was chosen. However, in similar experiments Hirschhorn and Rosecrans[47] did not observe generalization to cyclazocine and nalorphine in rats trained to distinguish morphine from saline.

Moreover, even the group of mixed agonists-antagonists seems to be heterogeneous. Hirschhorn[48] found that discriminated responding was readily acquired by reinforcing one lever after saline and the other upon treatment with nalorphine, cyclazocine, or pentazocine. However, while in the nalorphine/saline group, clear generalization was detected to pentazocine and cyclazocine, neither of them generalized to nalorphine.[48]

Two groups[43,47] reported on partial discrimination, i.e., applying pentazocine and cyclazocine instead of morphine the discrimination surpassed the chance level of 50% but never attained the critical 90%.

In a recent study[49] ethylketocyclazocine, pentazocine, and levallorphan produced rather weak stimulus effect in rats trained to dicriminate between cyclazocine and saline, while morphine and nalorphine showed strong stimulus efficacy.

The cross-generalization was studied also in relation of nonopiate drugs. The stimulus properties of LSD generalized partially to cyclazocine, but not to nalorphine, pentazocine, methadone, and meperidine.[47] The stimulus property of cyclazocine is surely not indistinguishable from that of the LSD, as at neither dose of cyclazocine was such a high percentage of correct responding observed than with LSD itself.[47] Consequently, the stimulus properties of the two drugs may be similar but they are surely not identical. Also in man the subjective effects elicited by LSD and cyclazocine are similar but not identical.[50]

The animal studies are consistent with the old clinical observations, according to the mixed opiate agonists-antagonists resemble in certain aspects the pure narcotic drugs, while in some other aspects the psychomimetics. Moreover also these data point to the multiplicity of the receptors mediating their effects.

Of course the generalization experiments were not restricted to narcotics and psychomimetics. Another old dispute has been practically solved by experiments of this type. It has been previously supposed that the narcotic drugs act at least in part via inhibition of pre- or post-synaptic dopaminergic receptors.[51] But morphine could not be successfully substituted by the dopamine receptor blocking haloperidol.[42] To be sure, Colpaert et al.[52] found that rats trained to discriminate apomorphine from saline choose the drug appropriate lever upon fentanyl treatment. However, they did not find stimulus generalization in the opposite direction and the cuing effect of apomorphine could be abolished by haloperidol but not by naloxone.[52] Thus the cuing effect of apomorphine might be explained by its weak and aspecific opiate agonist activity mediated by its σ-receptor agonist activity as supposed by Martin et al.,[23] but not by interaction with the postsynaptic dopaminergic receptors.

Among the other substances examined for stimulus generalization are compounds structurally related to the narcotics but lacking any opiate-like activity (dextrorphan, thebaine);[43] dopaminergic receptor blockers (chlorpromazine, azaperon);[43,53] presynaptic α-adrenergic receptor stimulants (clonidine);[53] anticholinergic agents (scopolamine, dexetimide);[54] cholinerg stimulants (pilocarpine, physostigmine);[44,54] psychomimetics (mescaline, ketamine);[44] sedato-hypnotics (ethanol, pentobarbitone)[43] and antiinflammatory-antipyretic agents (aspirin, indomethacin, phenylbutazone, phenacetine, tolmetin).[45]

Of these "negative" findings, the data on clonidine[53] and amphetamine[43] deserve special attention. The subjective effects induced by amphetamine resemble those induced by morphine.[37] Thus it is probably not surprising that in experiments of Shannon and Holtzman[43] rats trained to discriminate between morphine and saline did not press consequently the saline correct lever upon amphetamine administration. They practically vacillated between the two levers just around the chance level, proving that the substances induced partly similar but partly different subjective effects in rats.[43] Clonidine represents another interesting case. This drug does not interact directly with the opiate receptors[55] but displays significant analgesic potency[56,57] and suppresses the opiate withdrawal symptoms (Chapter 9, Volume II).

C. Is It Possible to Suppress the Stimulus Properties of Narcotic Drugs by Modifying the Nonopiate Functions of the Brain?

Haloperidol did not suspend the fentanyl-saline discrimination in rats[58] and the discrimination of amphetamine from saline was not altered by either morphine or fentanyl.[59] However, the discriminative effect of amphetamine could be reversed by haloperidol.[59] Thus dopamine receptors are not involved in the subjective effects induced by opiates. Depletion of the brain noradrenaline content by α-methyl-p-tyrosine also had no significant effect on the morphine cue.[60] However, depletion of the serotonin content of the brain resulted in abolition of the morphine's discriminative effect.[60] This extremely interesting finding, however, was not confirmed by others.[61] Since serotonin has no affinity to the opiate receptors, serotonin depletion might influence only those actions of opiates and opioid peptides which are brought about indirectly, via activation of serotoninergic pathways. As it will be discussed later (Chapter 2, Volume I and Chapters 2, 3, 4, and 5, Volume II) many opiate effects are connected with activation of serotoninergic functions and stimulation of the serotoninergic mechanisms facilitates many opiate-specific actions. Thus, this controversy on the role of serotonin in the stimulus properties of opiates deserves further experimentation.

Where are the receptors mediating the discriminative effects of opiates located?

Data concerning this exciting question are rather scarce. The receptors mediating the stimulus properties of opioids must be located centrally since morphine, fentanyl, sufentanyl exhibit narcotic cuing potency upon i.c.v. administration just as on peripheral treatment.[62,63] However, in an interesting study of Shannon and Holtzman[62] these discriminative effects of morphine observed via i.c.v. administration could not be consistently reproduced by injecting it directly into the periaqueductal gray matter, lateral septum, or dorsomedial thalamus, i.e., via several brain areas known to be rich in opiate receptors (Chapter 1, Volume III). In this experiment different behavioral effects were observed depending on the site examined, but the level of discrimination never reached that obtained upon i.c.v. treatment.[62] The appearance of behavioral symptoms excludes the possibility that in this experiment the doses applied were too small.[62] Thus the brain sites responsible for the stimulus properties must be elsewhere. In another study,[64] however, injection of morphine into the periaqueductal gray produced stimulus control similar to that observed upon peripheral application. Moreover, naloxone injection into this area effectively blocked the discriminative effect of peripherally injected morphine.[64] The only data[65] concerning the raphe complex are positive, i.e., rats trained to discriminate s.c. administered fentanyl from saline chose the drug lever if the former was injected into the nucleus raphe magnus.

Thus the experimental examination of stimulus control properties of narcotics is an extremely sensitive method for the in vivo detection, delination, and classification of opiate receptor mechanisms. Surely, the inherent possibilities provided by this method are far from being exhausted.

IV. ANALGESIC EFFECT

Among the characteristic in vivo actions of opioids analgesia is discussed only on the third place, since this effect contrary to any expectation seems to be a less specific peculiarity than the antagonizability by naloxone and the stimulus property. The antinociceptive action is regarded to be less specific for several reasons.

If opiates are administered peripherally or into the liquor space, i.e., not directly into several specific areas of the brain, the antinociceptive action appears always in the company of several other motor and autonomic symptoms as only one of the many manifestations of a complex behavioral pattern.

In some tests claimed to be specific for opiates, not only the morphine-like compounds are active.

Since the results of analgesia testing depends to a considerable extent on the assay used, the specificity of methods applied in measurement of antinociceptive action deserves special attention.

Of the methods used to quantify the antinociceptive effects of opioids probably the oldest one may be regarded as the most specific even today. The rat tail flick test, where radiant heat is the painful stimulus and the pain threshold is measured by the latency of tail withdrawal, was introduced by D'Amour and Smith[66] in 1941.

The so-called tail immersion test may probably be regarded as a modified version of tail flick assay.[67] In this case a water bath of constant temperature is used as painful stimulus. Just as in the tail flick,[66] also in this assay there is a very good correlation between the experimentally measured and clinically found relative potencies of narcotic drugs. The main advantage of this method with regard to the classical tail flick test lies in the adjustability of heat stimulus. Moreover using not too hot water, the permanent tissue damage can be avoided.

These tests are selective for morphine-like narcotic drugs. Neither other types of psychotropic drugs nor aspirin-like anti-inflammatory agents are active in these tests.[66,68] The well-known activity of some centrally active drugs, such as amphetamine,[68] clonidine,[56,57] and substance P[68] reflect their ability to excite, even if indirectly, the opiate receptors. A further advantage of the method is that the tail flick movement is a strictly spinal reflex, which also can be elicited in spinal preparation.[70] Thus using the tail flick test in spinal animals, the action at spinal level may be separately examined.

Probably both methods can be derived from that introduced by Haffner[71] in 1929. According to this procedure the root of the tail is gripped with artery forceps to produce a squeak. Of course the response of the animals to this noxious stimulation does not consist merely of vocalization. In many cases the pain induced vocalization and motor responses (struggle) are affected differentially,[72] which renders the test unreliable. Moreover, by using this test, the intensity of painful mechanical stimulation is neither quantified nor adequately controlled.

A further way of stimulation has been introduced in the present form by Grewal.[73]

During this procedure electric shocks are applied to the mouse tail and a squeak response is taken as pain reaction. An essential modification of this method has been introduced by Paalzow[56,57,74,75] and others determining the threshold of vocalization after-discharge. Upon standardized electrical stimulation of the rat tail three different consecutive responses can be seen; first a motor response mediated spinally, then a vocalization response mediated by structures in the medulla oblongata, and finaly a vocalization after-discharge (vocalization after withdrawal of the stimulus).[54,57,74,75] The latter response is considered to be a reaction mediated by brain structures involved in the emotinal component of pain reaction, such as thalamus, hypothalamus, rhinencephalon.[56,57,74,75] It is probably due to the complex nature of biological response that this method is not very widely used. Moreover, it is always questionable whether the electric stimulation represents a natural way of painful stimulation. In several laboratories the electric footshock induced "flinch" and "jump" responses are recorded as motor manifesatations of pain sensation.[76]

Of the other methods connected with electrical stimulation as pain source, the shock titration method deserves special attention.[77,78] In this paradigm electric shocks of gradually increasing intensity are applied to the footpads of rats or monkeys. Pressing down a lever decreases the intensity of current by an equal amount with each response. Thus, appearance or disappearance of responding indicate decrease or increase in the pain threshold.[77,78]

The main advantage of these methods is that the stimulus strength may be easily regulated and quantitated. However, electric current is not a natural way of painful stimulation and it is not clarified that actually which receptors are activated by it. Nevertheless, the vocalization in itself refers to the aversive character of this reaction.

Uncomparably less specific are two further tests used widely for measurement of analgesic action: the hot plate test,[79] where the latency time of licking of the limbs is measured upon placing the animals on a heated metal sheet and the writhing test,[80-83] where the characteristic movement and posture is recorded in mice treated intraperitoneally with acetic acid or phenylquinone. In both cases rather characteristic and complex motor reflexes are recorded, where the movements of the trunk, the lower and upper extremities and head are coordinated. Obviously both reflexes are organized far beyond the spinal level, thus they may be influenced by many higher brain structures primarily not related to the nociceptive reactions.

In the "inflamed foot" test,[84] i.e., measuring the threshold of mechanically induced pain on the swollen leg, obviously any drug with anti-inflammatory action may give positive results.

Thus in this book mainly those experiments will be discussed where tail flick or tail immersion tests have been used. Needless to emphasize, that in the above-mentioned assays the analgesia induced by opioids could be antagonized by naloxone-like drugs. In several cases the data obtained by nonspecific test are accepted as specific if the antagonizability by narcotic antagonist was sufficiently demonstrated. The specificity of tail flick and tail immersion tests is proved also by the steep dose-response curves recorded by them in most studies concerned. Conversely, the nonspecific, indirect nature of eventual effects displayed by nonopiates is reflected by their rather flat dose-effect curves. In this connection Sewell and Spencer's works[68] are worth mentioning, who, using the tail immersion test, compared the effects of pure narcotic analgesics with those of mixed agonist-antagonists. The pure narcotics (morphine, etorphine, meperidine) displayed characteristically steep and parallel log. Dose-response curves, while the partial agonists (cyclazocine, nalorphine, pentazocine) also produced parallel dose-response plots but they were rather flat.[68,85] Even naloxone, the "pure" antagonist showed a certain weak agonist action with very shallow dose-response plot.[68] Otherwise the data on the alleged agonist activity of naloxone are rather conflicting; they will be reviewed in Chapter 3, Volume II.

The question arises: which brain areas mediate the antinociceptive effects of opiates? Their sites of action may be examined by their local microinjections and also by drug interaction experiments. These experiments will be surveyed later (Chapter 2, Volume I). In this chapter only the data obtained by brain lesions are shortly discussed.

The primary site of their action must be within the brain, since transsection of the spinal cord was repeatedly demonstrated to drastically diminish the analgesic potencies of opiates.[86-89] However, in all these studies a certain residual activity was left.[86-89] This residual activity could be abolished by naloxone.[87] These experiments show the importance of descending spinopetal pathways in bringing about the analgesic action, but also prove the existence of intraspinal opiate mechanisms.

In this context the data of Satoh and Takagi[90] are especially interesting. These authors showed that the splanchnic afferent impulses recorded from ventro-lateral funiculus of cats were suppressed by morphine and this inhibition could be reversed by transsection of the spinal cord or by low medullary transsection but if severing the brain stem at higher level, the effect of morphine was retained.[90] Such lesions also suppress the inhibitory effects upon dorsal horn cell activities observed by electrical stimulation of nucleus raphe magnus.[91-93]

Also these results suggest that the analgesic effect of morphine is mediated mainly through facilitatory effect on the lower brain stem, on neurones exerting inhibitory influence on the spinal sensory transmission.

As shown by Martin et al.,[23] in dogs the spinal opiate receptors are not uniform, since the antagonizability by naloxone depends on the kind of agonist used. Unfortunately no data are available on the uniformity or multiplicity of opiate receptors in other species. Using, however, intrathecally administered morphine as agonist and naloxone (i.p.) as antagonist in rats the pA_2 values calculated were similar to those obtained via peripheral administration of the agonist[12] (in Chapter 2, Volume I). Thus in rats the spinal opiate receptors mediating the analgesic action might be similar to those in the brain. But the eventual similarity of spinal and supraspinal opiate receptors mediating the analgesic effect of morphine does not necessarily mean that they are connected with the same transmitters. Wilcox and Dewey[88] reported that in spinal mice contrary to intact ones atropine reduced the morphine analgesia. Thus the analgesic action of morphine at spinal level might be mediated at least in part by muscarinergic mechanisms, which is surely not the case in the higher brain structures (Chapter 2, Volume I).

Of the many brain regions studied, lesioning of only several ones resulted in significant modification of narcotic analgesia. First of all electrolytic lesion of the periaqueductal grey matter (PAG) and that of the raphe complex resulted in significant attenuation of opiate analgesia.[94-99] In these experiments the lesion induced loss of analgesic activity of morphine and related agents was not due to a decrease in the sensitivity to painful stimuli.

The raphe nuclei contain mainly serotoninergic neurones which project to various areas including cerebral cortex, hypothalamus, the limbic system, caudate nucleus, and spinal cord. Electrolytic lesion of these nuclei results in a marked and long-lasting decrease in the forebrain levels of serotonin and 5-hydroxyindoleacetic acid.[94,97-99,100] As the transection of the spinal cord or specifically the funiculus dorso-lateralis also results in inhibition of morphine analgesia,[96] first of all the descending spinal pathways may be implicated. Of course it cannot be excluded that this loss of analgesic potency is due to the destruction of ascending pathways connected with the raphe system. These data only underlie the critical importance of serotonin releasing neurones in opiate analgesia. All the more, since the effect of raphe lesions could be reversed by administration of serotonin and 5-hydroxytryptophan. Moreover, i.c.v. administration of 5,6-dihydroxytryptamine, a selective and irreversible depletor of brain serotonin also results in attenuation of morphine analgesia.[97,101,102] The only conflicting data were reported by Bläsig et al.[103] who did not find the effect of morphine attenuated either by electrolytic or by chemical lesions of the raphe complex.

Locus coeruleus is a further site in the brain stem, which is implicated in narcotic analgesia. Some authors observed diminution,[97,104,105] others an increment[106] or no effect[107] in the analgesic potency of morphine upon lesioning of this area. This effect might, however, be connected with alteration in pain sensitivity, which was found enhanced,[97] decreased,[105] or unaltered[106] following the lesion. As the destruction of these neurones results in a significant depletion of forebrain noradrenaline content, the question is connected with the contingent role of noradrenergic mechanisms in narcotic analgesia or with the supposed role of locus coeruleus in regulating the activity of raphe nucleus.[106]

Similarly conflicting reports have been published on the consequences of forebrain dopamine and noradrenaline depletion (see Chapter 2, Volume I). The main drawback of these experiments is that upon i.c.v. treatment very different structures are simultaneously depleted. In the experiments of Nakamura et al.[108] the attenuation of analgesia upon intrastriatal treatment with 6-hydroxydopamine could be reversed by 1-DOPA + a peripheral DOPA decarboxylase inhibitor (Ro 4-4602). Moreover, selective depletion of the noradrenaline in rat hypothalamus or medial forebrain bundle had opposite effect.[108]

Of the other brain areas implicated in morphine analgesia by the technique of brain lesioning two others deserve attention. Destruction of medial thalamus was reported to potentiate morphine analgesia in rats.[109] Others, however, found a reversal of morphine induced decrement in cortical acetylcholine release upon lesioning of the medial thalamus.[110] Recently, attenuation of morphine analgesia was reported upon destruction of the preoptic forebrain region.[104,111]

Thus raphe nuclei, PAG, and preoptic forebrain regions are the brain structures destruction which results in attenuation of narcotic analgesia. Since these areas are known to be interconnected with each other[104,107] the neurones needed for opiate analgesia appear as a common circuitry. Confirming the previous histological studies the existence of direct projection from different regions of the PAG into the raphe nuclei has recently been demonstrated by electrophysiological methods.[112]

V. BEHAVIORAL EFFECTS

The main reason of analyzing the motor effects induced by opioids is their inseparability from their "principal" effect, i.e., analgesia. Opiates elicit rather complex behavioral effects and just as in the cases of certain clinical syndromes generally not the symptoms themselves but their pattern is an easily recognizable entity. (The only exception is probably the tail erection, called Straub phenomenon, a peculiar sign which in itself refers to the excitation of opiate receptors. Unfortunately this symptom appears only at relatively high dose levels in several species of rodents and it is not easily quantifiable.) The behavioral effects are, however, very variable depending on the dose applied, the interval elapsed since drug administration, and on the species used.

The diversity of behavioral symptoms might also be attributed to opiate receptors. However, considering the similarity of pA_2 values of naloxone against the different morphine derivatives (Chapter 2, Volume I) also other possible explanations must be taken into account. Thus, considering that the presynaptically acting opiates modify the release of many transmitters in very different brain regions (Chapter 1, Volume II), this diversity is not surprising. Moreover, their actions may initiate compensatory mechanisms, which also confound the primary effects. Upon repeated application tolerance development, receptor supersensitivity, and other adaptive processes may modify the outcome of experiments. Nevertheless, applying them in equiactive dosage in the same species and taking into account the characteristic sequence of events, very similar behavioral symptoms will be detected giving whichever narcotic drug. These rather motley behavioral effects are worthy of detailed analysis for various reasons. First of all these can be antagonized by naloxone-like drugs, thus they reflect the in vivo activation of opiate receptors. One behavioral symptom, the Straub phenomenon, and the pattern of other behavioral changes are more characteristic than the analgesic effect itself. Finally a great part of our present knowledge on the in vivo mechanisms of opiate actions is connected with the pharmacological analysis of these effects.

In most pharmacological textbooks, upon discussing the behavioral actions of narcotic drugs, the species differences are emphasized. For the unexperienced observer the behavioral symptoms are surprisingly different; e.g., in rats deep sedation, areflexia, immobility, muscular rigidity, lack of reactions to certain external stimuli, loss of righting reflex, and tail-erection are the leading symptoms. In mice, however, the locomotor activity seems to be increased, the animals, though not reactive to painful stimuli, seem to be very irritable and discounting the tail-erection the increase in muscular tone is not very obvious. In cats the main signs are of excitatory nature such as restlessness, vocalization, mydriasis, piloerection, and sterotypy. But in dogs similarly to rats principally sedative effects can be observed.

A separate analysis of the individual effects and their sequence, and the thorough examination of dose-effect relationships reveals, that narcotics induce rather similar effects in different species. Only the relative contribution of separate symptoms to the whole picture is variable, depending on the species examined, dose applied, and time elapsed since treatment. Thus it is probably reasonable to discuss the behavioral items separately.

A. Changes in the Spontaneous Motor Activity

As mentioned above, morphine-like drugs elicit a characteristic mixture of depressant and excitatory effects. Thus, the question arises: which of them may be related to the analgesic effect? Several lines of experimental data indicate that in rats the sedative effects are primarily connected with the antinociceptive action.

In rats at very low dose levels morphine facilitates certain behavioral reactions such as grooming and locomotion without significant analgesia. At moderate and high dose levels the antinociception appears in the company of the well-known depressant effects such as decreased locomotor activity, catalepsy, increase in muscular tone, areflexia, etc.[113-117] Furthermore, it has been demonstrated that in rats the behavioral effects of opiates are biphasic, i.e., the locomotor depression is invariably followed by excitation. The higher the dose is applied, the later the secondary excitatory phase appears. But during this late phase of excitatory symptoms, no analgesia can be observed but hyperreflexia and probably a decrease of pain threshold.[114,116,118]

It is well known that upon repeated administration tolerance develops both to the analgesic and the accompanying depressant effects. As for the late excitatory effects no tolerance development can be observed. Actually this increase in activity appears progressively sooner and sooner if the opiate treatment is repeated.[113,119] Thus by gradual dissipation of depressant and analgesic actions stimulation comes to the front in a species in which narcotics are generally qualified as sedatives.

It is an everyday experience that narcotic antagonists inhibit not only the analgesic action of opiates but also the above-mentioned depressant effects. In relation of the second excitatory phase the data available are far from being unanimous.[114,116,120,121] Nevertheless, these data indicate that giving naloxone together or before the agonists both the initial depressant and the subsequent excitatory symptoms may be prevented. Giving naloxone after the dissipation of sedative phase, the excitation cannot be inhibited.[114,120,122]

It seems to be justified to conclude that the initial depressant effects may be regarded the specific behavioral manifestation of opiate receptor stimulation. The question arises: do these late effects, in many aspects of the opposite of the initial action, reflect rebound or compensatory mechanisms? This is tempting speculation in view of the general transmitter release inhibiting actions of opioids (Chapter 1, Volume I and Chapter 1, Volume II). It is easy to imagine that during the suppression of transmitter release the corresponding postsynaptic receptors may become supersensitive, resulting in hyperactivity upon termination of the primary drug effect. According to such a hypothesis all narcotic drugs are expected to elicit the same sequence of events.

However, there are several morphine congeners such as etrophine, levorphanol, and methadone, which contrary to morphine elicit only analgesia and sedative effects upon administering them into the periaqueductal gray matter but no excitation,[123] an interesting finding, which prompted Jacquet[124] to postulate the existence of a dual mechanism in mediation of opiates' effects. She[124] supposed that "classical" opiate receptors elicited analgesia and the sedative effects and another group of neurones were responsible for the excitatory actions.

Conversely, some morphine derivatives were shown to have diminished affinity to the opiate receptors with concomitant enhancement of excitatory potency.[125] Introducing glucuronide, acetyl, or sulfate groups in positions 3 and/or 6 of the morphine molecule, an inverse relationship was found between receptor binding and motor excitatory potencies.[125]

Numerous experimental findings show that the two phases of behavioral activity may be modified differentially by pharmacological manipulation. Thus Vasko and Domio[114] observed an initial decrease and a subsequent increase in acetylcholine utilization in the rat brain at the time of decreased and increased spontaneous locomotion, respectively. Upon repeated morphine treatment tolerance developed to its depressant effect on locomotor activity and to the decrease in the acetylcholine utilization as well.[114] The morphine-induced locomotor depression may be inhibited by p-chlorophenylalanine[119,126] infering the importance of serotoninergic mechanisms in the CNS depressant actions of narcotic drugs. As for the morphine-induced locomotor facilitation, it could be inhibited by atropine and scopolamine,[120] indicating the participation of parasympathomimetic mechanisms in this phenomenon. And a similar inhibition of this phase was observed upon depletion of both catecholamines by α-methyl-p-tyrosine or that of noradrenaline by FLA-63 or U-14624 or by blocking the dopaminergic or α-adrenergic receptors by pimozide, aceperon, and phenoxybenzamine.[118,120,127] Thus, both cholinergic and catecholaminergic mechanisms may be involved in the excitatory action.

Consequently, the initial behavioral depression might be related to serotoninergic and the subsequent excitation to cholinergic and catecholaminergic mechanisms. But before accepting this oversimplified explanation, some other alternatives must be considered. Jacquet and Lajtha[123] observed concomitant hyperreactivity to auditory, visual, or light tactile stimulation and hyporeactivity to painful stimuli, i.e., analgesia if morphine was injected into the periaqueductal gray. In another study the same group observed either analgesia or hyperalgesia depending on site and dose of intracerebrally given morphine,[128] while with some other narcotics only depressant actions could be elicited.[123] Jacquet[124] actually supposed that morphine acted on two different receptor populations in the brain, the first being naloxone sensitive endorphin receptors the others naloxone insensitive ACTH receptors mediating the excitatory effects of the drug and the opiate abstinence syndrome. The possibility arises that not two successive phases of a unitary process are expressed by the opiate-induced behavioral depression then excitation but the simultaneous activation of two different even antagonistic groups of neurones. It is not a pure academic issue whether the diversity of behavioral effects represent two phases of a uniform process or the simultaneous activation of two opposite mechanisms. Namely in the latter case the multiplicity (at least the duplicity) of opiate receptors must be supposed, in the former case, however, a unitary opiate receptor population and the secondary activation of nonopiate mechanisms might be supposed.

Domino et al.[115] reported that morphine in a very low subanalgesic dose caused behavioral excitation. This excitatory effect at subanalgesic, i.e., subsedative dose level reminds of the similarly "paradox" excitatory action of minor tranquilizers also at subthreshold dose level. (For literature see the corresponding pharmacological textbooks.) It is tempting to speculate that in both cases the behavior reflects a certain disinhibitory effect due to their anxiolytic action. If it is true, separate receptor populations should be supposed in mediation of the analgesic and certain behavioral effects.

The whole matter becomes even more complicated if the species differences are also taken into consideration. Besides rats there are some other species (e.g., dog) in which the depressant effects prevail,[116] but the late excitatory effect in this species (unpublished observations in our laboratory) are generally not commented. Other laboratory species such as cats or mice are known to react with predominantly excitatory effects to opiate administration. The symptoms observable in cats are summarily called "feline mania"[116,129] and those in mice "running fit".[122,130,131] "Feline mania" refers to vigorous motor excitation with periods of rage-like behavior and the corresponding autonomic manifestation of increased sympathetic activity (mydriasis, exophthalmos, piloerection), while "running fit" in mice means enhanced locomotor activity with a tendency to circle around.[116,122]

In cats applying opiates in low dosage diminished locomotion and stationary posture can be observed immediately after treatment.[129,132-134]

The much discussed excitatory symptoms appear only later and at higher dose levels. The manic response induced by morphine can be prevented by depletion of the brain catecholamines (with reserpine or tetrabenazine) or by dopaminergic receptor blockers (such as haloperidol or chlorpromazine).[129] On the other hand α- and β-adrenerg receptor blocking agents (phenoxybenzamine and propranolol) did not antagonize the feline mania.[129] Thus, contrary to the oversimplified descriptions (i.e., catalepsy in rats but motor excitation in cats), the more detailed examination reveals a pattern rather similar to that observed in rats: sedation at low dose level immediately after treatment then excitation. It is characteristic both for cats and rats that the animals do not react to painful stimuli but they are hyperresponsive to visual, tactile, and auditory stimuli (see above). Probably the peculiar autonomic signs of increased sympathetic activity (first of all piloerection and mydriasis) contribute to the subjective impression that the morphine treated cats contrary to rats are not sedated but activated. Nevertheless, as discussed above, catecholaminergic mechanisms mediate the excitatory actions of opiates in this species too.

As for mice, so far almost exclusively the excitatory effects have been emphasized.[116,122,130,131,135-137] These animals might represent the only laboratory species where the stimulant action prevails without a primary depressant phase. But recently Hecht and Schiørring[138] reported that morphine elicited transient decrease in spontaneous activity in mice. This finding has been confirmed in our laboratory.[139] In our experiments morphine elicited a dose-related inhibition of locomotion lasting for 15 to 30 min, which was then followed by excitation[139] (see Table 2). The apparent

Table 2
EFFECTS OF MORPHINE AND D-MET, PRO-EA ON SPONTANEOUS MOTILITY IN MICE[a]

	Treatment	Dose mg/kg s.c.	0-15	15-30	30-45	45-60	60-75	75-90	90-105	105-120	Total	No. of groups used	Statistically compared groups
I.	Saline	—	494±79	419±124	204±80	145±60	137±58	57±34	53±32	64±27	1574±412	13	—
II.	Morphine	1	457±71	313±65	353±68	250±79	385[b]±105	172±51	175±77	116±57	1717±246	12	II.vs.I.
III.	D-Met, Pro-EA	1	328±41	399±68	326±55	242±65	133±46	86±40	100±27	102±73	1717±246	16	III.vs.I.
IV.	Saline	—	634±74	314±70	288±70	256±68	161±46	136±41	123±42	111±28	2029±352	15	—
V.	Morphine	3	189[c]±47	236±45	249±56	318±53	461[d]±78	513[c]±106	405[c]±87	334[c]±77	2666±355	13	V.vs.IV.
VI.	D-Met, Pro-EA	3	245[c]±27	205±61	313±44	224±27	182±49	121±31	93±38	111±48	1494±208	15	VI.vs.IV.
VII.	Saline	—	632±128	272±94	135±62	111±48	220±77	70±27	40±19	152±86	1633±307	12	—
VIII.	Morphine	10	194[c]±47	191±43	291±60	276[b]±51	344±76	367[d]±58	356[c]±73	535[c]±72	2522±72	15	VIII.vs.VII.
IX.	D-Met, Pro-EA	10	171[d]±34	69[b]±17	199[d]±47	297[c]±36	324±54	258[c]±52	153[c]±40	111±34	1581±204	15	IX.vs. VII.

[a] Locomotor activity was measured in actometers equipped with photocells which operated electromechanical counters. Mice were examined in groups of three animals.
[b] $p < 0.05$.
[c] $p < 0.01$.
[d] $p < 0.001$.

contradiction of these data is probably of methodological origin. In the earlier experiments[122,130,131,135,137] the locomotion was recorded mostly cummulatively, while in the latter[138,139] the subsequent time periods were evaluated separately. Thus, we have to suppose that in the earlier experiments the initial relatively short-lasting phase of locomotor depression was obscured by the next phase of hyperactivity. Nevertheless, this syndrome of behavioral excitation can easily be antagonized by naloxone and nalorphine and upon repeated administration tolerance develops to this action.[131,137] Moreover, the running fit can be elicited only by the biologically active enantiomers of opiates.[131] The slopes of the dose-response curves for analgesia and "running fit" were found to be identical.[131] Thus, this peculiar effect meets the main criteria of opiate specific actions, which cannot be said of the opiate-induced hyperactivity in rats (see above). Otherwise, the brain transmitters seem to play partially similar partially different roles in mediation of opiate induced hyperactivity in rats and mice, respectively. Thus, just as the hypermotility in rats and cats, the levorphanol induced running fit could be prevented by α-methyl-p-tyrosine and reserpine,[135] proving the importance of catecholamines in this phenomenon. However, in contrast with the observations made in rats, the locomotor activation in mice can also be inhibited by increasing the parasympathetic tone via administration of physostigmine.[130] The morphine-induced excitation can also be prevented by diethyldithiocarbamate as well, i.e., by inhibition of dopamine β-hydroxylase.[140] Of the catecholamines, epinephrine (or norepinephrine) might play a crucial role. The dependence of morphine induced running fit on the noradrenergic system is further substantiated by the ability of clonidine to reverse the inhibition caused by inhibition of dopamine-β-hydroxylase.[140] On the other hand, pretreating the mice with p-chloro-phenylalanine, the levorphanol induced running fit was not found modified.[135] Thus, the brain serotonin plays no significant role in the hyperactivity induced by opiates in this species either.

Several other findings also show that analgesia and running fit are two different, independent even mutually antagonistic phenomena. Thus a negative correlation has been found between morphine- and heroin-induced running activity and analgesia measured by the hot plate test if different inbred strains were compared.[136] Mice specifically sensitive to the locomotor stimulant action of levorphanol show enhanced motor response to amphetamine as well.[141] These animals displayed a relatively weak analgesic response in the hot plate test.[141] Thus either the motor exaltation disturbed the correct measurement of antinociceptive effect, which may be counted with if using this test so sensitive to unspecific interference, or the locomotor activation and antinociception are mutually antagonistic effects. Interestingly of the numerous morphine derivatives examined in this context only one compound, ethylketazocine (WIN 35,197-2) has been shown to decrease the locomotor activity in mice.[142] Upon its repeated administration the locomotor depressant activity diminished, i.e., tolerance developed to its depressant effect.[142] Thus in mice tolerance develops to the locomotion modifying effects of opiates whether hypermotility or hypoactivity is induced by them. Considering that ethylketazocine is supposed to be a specific agonist of the so called $\varkappa$-receptors,[23,24] the above-discussed locomotor stimulant action of classical morphine derivatives might be specifically related to the μ-receptors. Of course, further experiments are needed to draw such a conclusion, all the more since the μ- and $\varkappa$-receptors were separated on the basis of experiments made in dogs and not in mice.

Thus in mice, contrary to rats, the locomotor stimulation might be rather closely related to the opiate receptors and it does not seem to be only a secondary rebound-like effect. But even in this species partially different mechanisms seem to be involved in the mediation of these phenomena, which under special conditions appear as mutually antagonistic symptoms.

Otherwise, of the many opiate-induced behavioral signs in mice and to a lesser extent in rats probably the tail-erection (Straub phenomenon) is the most characteristic and the simplest symptom to recognize.[116] Unfortunately this phenomenon has not been analyzed separately from the other behavioral parameters.

B. Catalepsy, Rigidity

Beside analgesia and decreased spontaneous locomotion, catalepsy is the most characteristic behavioral symptom observable in rats treated with relatively high doses of narcotic analgesics. It is most conspicuous in rats, in dogs, and in monkeys but in the latter species no specific tests have been developed to quantify this symptom.[116] Discounting one report,[143] no catalepsy has been observed in mice and in cats. Only Beecham and Handley[143] observed a weak and hardly detectable cataleptogenic effect upon morphine, codeine, and pethidine treatment. *Per definitionem* catalepsy means the maintenance of abnormal posture, which may be standardized for appropriate scoring in rats.[144,145] To the experienced observer this symptom seems to be a consequence of other behavioral events such as lack of spontaneous movements, unresponsiveness to certain external stimuli, and increased muscular tone.

Catalepsy may be elicited not only with opioids but also by cholinergic and neuroleptic agents.[144,146] However, the behavioral state of immobility induced by morphine-like agents and that elicited by neuroleptics are not the same. They are different qualitatively and they also may be differentiated on the basis of brain lesion and drug interaction techniques.

Thus, contrary to the catalepsy induced by neuroleptics, the immobility observed upon administration of morphine congeners is characterized by extreme rigidity. Consequently, it seems to be justified to differentiate between catalepsy (after neuroleptics) and catatonia (after narcotics).[146,147] The latter produce "lead pipe" rigidity of the trunk's and limbs' musculature, loss of righting reflex, and a total lack of any spontaneous motor activity[146,147] (see also Chapter 3, Volume III). Animals treated with chlorpromazine-like agents also are immobile but their muscular tone is not so extremely high and they show normal righting reflex, i.e., these animals cannot be placed in supine position. The different nature of immobility induced by the two classes of drugs is further confirmed by another so far unexplained observation.[143] Applying a strong pressure to the mouse or rat tail (placing an artery clip with or without jaws on it) enhances the opiate-induced catalepsy while abolishing the similar effect of neuroleptics.[143]

Opiate antagonists completely reverse the morphine-induced catatonia,[46] but atropine and scopolamine do not.[146,148] The latter ability to antagonize the cataleptogenic effect of neuroleptics is a well-known phenomenon. Thus, catalepsy is an opiate specific effect, which is mediated or initiated by neurones not identical with those responsible for the similar action of neuroleptics. The chlorpromazine-like drugs act via blockade of dopaminergic receptors, their specific antagonists are the dopaminergic stimulants such as 1-DOPA, apomorphine, and amphetamine. But these dopamine receptor stimulants also reverse the cataleptogenic action of narcotic drugs.[149] In this context it is worth mentioning that morphine and methadone raise the concentration of homovanillic acid in the striatum.[140,149] Thus, narcotic agonists elicit catalepsy with concomitant increase in the striatal dopamine turnover[140,149] and both effects can be reversed by naloxone. On the other hand, the cataleptogenic effect of neuroleptics is also accompanied by increase in the striatal turnover but their metabolic effect is specifically sensitive to dopaminerg agonists (see any pharmacological textbook for references). Though initiated by different receptors, the cataleptogenic actions of narcotics and neuroleptics as well, seem to be mediated by striatal mechanisms. Considering, however, the outcome of brain lesion experiments, even this assumption must be rejected.

As reported by Costall and Naylor[146] the cataleptic state induced by neuroleptics was reduced by lesions of the caudate-putamen or globus pallidus but that elicited by narcotic analgesics was unchanged or rather enhanced. And similarly, lesions placed into the lateral hypothalamus to destroy the dopaminergic input to the extrapyramidal and mesolimbic brain areas were found to reduce or abolish the cataleptogenic effect of haloperidol, while the similar action of morphine was found significantly potentiated by the same lesion.[146] However, electrolytic coagulation of the nucleus amygdaloideus centralis was shown to have opposite effect, i.e., this lesion completely blocked the cataleptogenic action of morphine, while that of haloperidol was only slightly inhibited.[146] In another study the same authors reported on the loss of morphine's ability to induce catalepsy in rats upon electrolytic destruction of medial and dorsal raphe nuclei.[150]

C. Stereotypy

As described in the pharmacological textbooks, in all experimental species the opiate-induced excitatory behavioral symptoms have a certain stereotyped character (i.e., circling and turning in mice, gnawing and biting in rats, repetitive jerking movements of the head, trunk, and limbs in cats and dogs).[116] As it will be discussed in Chapter 2, Volume II these stereotyped movements and the other manifestations of hyperactivity are due to a smaller or greater extent to the epileptic properties of opiates and especially of opioid peptides. Since the epileptic effects of opioids can be prevented by naloxone-like drugs just as the stimulant actions as discussed above, all these actions are directly or indirectly related to the opiate receptors.

Nevertheless, the opiate-induced inhibitory and motor stimulant effects are more or less antagonistic. Upon repeated morphine administration the initial behavioral depression and catalepsy are replaced step by step by general motor stimulation and stereotypy.[117,119] Just as in the case of catalepsy, the involvement of striatal mechanisms in generation of stereotypy must be counted with but in this case the functional balance being shifted in the opposite direction. As stereotypy in general and the apomorphine or amphetamine-induced stereotypy specifically are regarded the most characteristic behavioral manifestation of dopaminergic stimulation,[151,152] the results of corresponding drug interaction studies are not easy to interpret. Fog[117] reported on an antagonism between morphine and amphetamine in induction of stereotypy, and others[153] found a similar antagonism between apomorphine and morphine but the inhibition of stereotypy was not unambiguous in this[117] and in other studies[154,155] either. In our experiments (unpublished) morphine inhibited the apomorphine and amphetamine induced stereotyped behavior only in fairly high doses (10 to 30 mg/kg s.c.), which were sufficient to induce catalepsy. Thus the inhibition by morphine of stereotypy induced by dopamine receptor stimulants seemed to be an unspecific effect due more to the general motor debilitation than to specific interaction at receptor level. Bergmann et al.,[156] however, found the morphine induced gnawing potentiated by systemically administered amphetamine if the former was implanted into the ventral thalamus. In their experiments both apomorphine and morphine induced gnawing was inhibited by chlorpromazine, haloperidol, and pimozide but only the latter's effect by α-methyltyrosine or α-methyl-DOPA.[156] Thus the authors supposed that action of morphine is mediated by catecholamine release.[156] Potentiation by morphine of the nigro-striatal dopaminergic system also has been induced by Iwamoto and Way,[157,158] who reported on contralateral circling in rats upon unilateral intranigral administration of different opiates. As for the role of brain serotonin, Bergmann et al.[126] observed the potentiation by p-chlorophenylalanine of morphine induced gnawing and Costall and Naylor obtained the same effect by lesioning the medial or dorsal raphe nuclei.[150] Thus, inhibition of serotoninerg functions in the brain potentiates the narcotic-induced stereotypy and inhibits their cataleptogenic action.

Bergmann et al.[126] and Costall and Naylor[150] supposed that morphine and the related analgesic drugs exerted a dual effect: stimulation of stereotypy via catecholaminergic mechanisms and inhibition by serotoninergic mechanism. Or in a broader sense, induction of different categories of behavioral activation by the catecholamines, while induction of analgesia, sedation, and catalepsy by the serotonin neurones of the raphe nuclei.

VI. DRUG INTERACTION STUDIES

A. Drugs Affecting Brain Catecholamines and Serotonin

There is a considerable bulk of evidence suggesting that brain catecholamines and serotonin are involved in opiate analgesia. Moreover, the indirect nature of opiate actions has been amply demonstrated in vitro. One of the most important conclusions drawn from the experiments with the isolated organ preparations (mouse vas deferens, guinea pig ileum, etc.) was that opiates exert their specific effects presynaptically via inhibition of release of different neurotransmitters (Chapter 1, Volume I and Chapter 1, Volume II). Of course these conclusions are not necessarily valid also concerning the living organism. If these assumptions were true, depletion of the brain transmitters, or facilitation of their release, blockade, or activation of the corresponding postsynaptic receptors should modify the opiate analgesia in the direction predicted on the basis of the in vitro experiments. Under in vitro conditions the liberation of a great many transmitters might be found altered if the opiates are added to the organ preparations in sufficiently high concentrations. Of course only by the aid of carefully designed in vivo experiments can it be established which of these actions determine the netto action in the living organism.

Reducing the brain levels of biogenic amines by reserpine or tetrabenazine, i.e., by agents depleting both serotonin and catecholamines the analgesic effect of morphine congeners were found lost[154,159-165] (see also Table 3). However, in several studies reserpine was found to increase their analgesic potency[164,166-171] (see also Table 3). In the latter studies obviously the sedative effect of reserpine, resulting in lengthening of the latency time of different reflexive motor reactions, led to the erroneous conclusion on synergism between the two types of sedative agents.

One group (Ross and Ashford[169]) reported on potentiation in the mouse hot plate test but antagonism in the mouse tail clip procedure just as Tardos and Jobbágyi[171] did who observed synergism in the mouse hot plate test but no interaction in the rat tail flick test, applying reserpine 30 min prior to morphine.

A further possible explanation of these contradictions is probably connected with the time interval between application of reserpine and morphine. If this interval is too short, the catecholaminergic or serotoninergic tone may even be increased due to the reserpine elicited initial liberation of biogenic amines. Applying longer intervals between the two treatments, i.e., injecting morphine after depletion of the brain biogenic amines, the potency of morphine was greatly reduced.[164] This misunderstanding arose during the late 1950s and early 1960s and shows how faulty conclusions may be drawn if the in vivo experiments are not carefully planned (in many cases it is simply due to the lack of knowledge concerning the biochemical mode of action).

Obviously this effect of reserpine-like agents do not prove á priori the exclusive role of catecholamines and serotonin, as the existence of other yet unknown transmitters also depleted cannot be excluded. Considering, however, that by simultaneous administration of l-DOPA and 5-hydroxytryptophan, i.e., by restoration of normal brain levels of biogenic amines, the analgesic potency of narcotics can be restored, the causal relationship may probably be accepted.[160]

Table 3
EFFECTS OF PRETREATMENT WITH RESERPINE-LIKE AGENTS ON ANALGESIA INDUCED BY OPIATES

Pretreatment	Treatment	Species	Assay	Effect	Ref.
Reserpine	Morphine	Mouse	Tail flick	Antagonism	172
Reserpine	Morphine	Mouse	Hot plate	Potentiation	171
Reserpine	Morphine	Rat	Tail flick	No effect	171
Reserpine	Meperidine	Mouse	Hot plate	Potentiation	171
Reserpine	Meperidine	Rat	Tail flick	No effect	171
Reserpine	Morphine	Mouse	Tail pinch	Antagonism	162
Reserpine + iproniazid	Morphine	Mouse	Tail pinch	Reversal of the antagonistic effect of reserpine	162
Reserpine	Morphine	Mouse	Tail flick	Antagonism	173
Reserpine	Codeine	Mouse	Tail flick	Antagonism	173
Reserpine	Meperidine	Mouse	Tail flick	Antagonism	173
Reserpine + iproniazid	Morphine	Mouse	Tail flick	Reversal of the antagonistic effect of reserpine	173
Reserpine	Morphine	Mouse	Hot plate	Potentiation	166
Reserpine	Morphine	Mouse	Tail flick	Potentiatin	170
Reserpine	Morphine	Rat	Hot plate	Potentiation	170
Reserpine	Morphine	Mouse	Hot plate	Antagonism	174
Reserpine	Morphine	Rat	Tail flick	Antagonism	174
Reserpine + iproniazid	Morphine	Rat	Tail flick	Reversal of the antagonistic effect of reserpine	174
Reserpine	Morphine	Mouse	Tail pinch	Antagonism	160
Tetrabenazine	Morphine	Mouse	Tail pinch	Antagonism	160
Reserpine + l-DOPA	Morphine	Mouse	Tail pinch	Reversal of the antagonistic effect of reserpine	160
Tetrabenazine + l-DOPA	Morphine	Mouse	Tail pinch	Reversal of the antagonistic effect of reserpine	160
Reserpine + 5-hydroxytryptophan	Morphine	Mouse	Tail pinch	Reversal of the antagonistic effect of reserpine	160
Reserpine	Morphine	Mouse	Writhing	Antagonism	175
Reserpine	Morphine	Mouse	Tail clip	Antagonism	169
Reserpine	Morphine	Mouse	Hot plate	Potentiation	169
Reserpine	Morphine	Rabbit	Tooth pulp stimulation	Antagonism	159, 176
Reserpine	Morphine	Mouse	Hot plate	Antagonism	159, 176
Reserpine	Morphine	Rat	Tail pinch	Antagonism	154
Reserpine simultaneously with	Morphine	Mouse	Tail flick	Potentiation	164
Reserpine 16 hours prior to	Morphine	Mouse	Tail flick	Antagonism	164
Oxypertine	Morphine	Mouse	Tail flick	Antagonism	164
Reserpine	Morphine	Mouse	Tail flick	Antagonism	177
Tetrabenazine	Morphine	Mouse	Tail flick	Antagonism	177
Oxypertine	Morphine	Mouse	Tail fick	Antagonism	177
Reserpine	Morphine	Mouse	Writhing	Antagonism	178
Reserpine	Morphine	Mouse	Hot plate	Antagonism	178
Reserpine	Morphine	Rat	Inflamed foot	Antagonism	179
Reserpine + serotonin	Morphine	Rat	Inflamed foot	Reversal of the antagonistic effect of reserpine	179

Table 3 (continued)
EFFECTS OF PRETREATMENT WITH RESERPINE-LIKE AGENTS ON ANALGESIA INDUCED BY OPIATES

Pretreatment	Treatment	Species	Assay	Effect	Ref.
Reserpine	Morphine	Mouse	Hot plate	Antagonism	161
Reserpine + 5-hydroxytryptophan	Morphine	Mouse	Hot plate	Reversal of the antagonistic effect of reserpine	161
Reserpine	Morphine	Intact rat	Tail flick	Antagonism	165
Reserpine	Morphine	Spinal rat	Tail flick	No effect	165
Reserpine	Meperidine	Intact rat	Tail flick	Antagonism	165
Reserpine	Morphine	Mouse	Hot plate	Potentiation	167

Table 4
EFFECTS OF MONOAMINE OXIDASE INHIBITORS ON ANALGESIA INDUCED BY OPIATES

Pretreatment	Treatment	Species	Assay	Effect	Ref.
Iproniazid	Morphine	Mouse	Tail flick	No effect	173
Iproniazid	Morphine	Rat	Tail pinch	Antagonism	154
Tranylcypromine	Morphine	Mouse	Tail flick	Potentiation	177
Pargyline	Morphine	Mouse	Tail flick	No effect	177
Iproniazid	Morphine	Mouse	Hot plate	Potentiation	161
Deprenyl	Morphine	Mouse	Tail flick	Potentiation	181
Pargyline	Morphine	Mouse	Tail flick	Potentiation	181
Clorgyline	Morphine	Mouse	Tail flick	No effect	181
Phenelzine	Morphine	Mouse	Hot plate	Antagonism	180
Phenelzine	Meperidine	Mouse	Hot plate	Potentiation	180

Conversely, the antinociceptive action of narcotic drugs was enhanced or inhibited by reserpine and was reversed by monoamine oxidase inhibitors, i.e., by increasing the brain concentration of biogenic amines[161,173,177,180,181] (see also Table 4). Since 1-DOPA or 5-hydroxytryptophan given alone did not completely restore the analgesic potency of morphine lost upon reserpine treatment,[160] both serotonin and catecholamines should be involved in morphine analgesia. However, determining the relative importance of individual biogenic amines is more difficult.

The importance of serotonin is proven unambiguously, namely by pretreatment with p-chloro-phenylalanine (PCPA), a specific depletor of brain serotonin.[182] The analgesic effect of morphine and that of other opiates were to a large degree but not completely antagonized both in mice and rats[161,167,178,183-187] (Table 5). Irreversible depletion of brain serotonin by i.c.v. administered 5,6-dihydroxytryptamine also resulted in significant attenuation of morphine analgesia.[101] Moreover, serotonin given i.c.v. or its precursor 5-hydroxytryptophan and drugs increasing the central serotoninergic functions potentiated the analgesic effect of morphine and reversed the inhibitory action of PCPA or reserpine[160,161,162,164,167,173,177,179,188-190] (Tables 3,5,6,7). When injecting PCPA 3 hr prior to morphine no antagonism of opiate analgesia and no serotonin depletion were observed.[183] Thus, the inhibitory action of PCPA correlates well with its serotonin depleting action, i.e., its effect reflects the inhibition of brain serotoninergic functions and it is not due to any direct interaction between PCPA and narcotic drugs.[183] 5-hydroxytryptophan potentiates not only the morphine-induced analgesia but also its cataleptogenic and sedative effects.[205] The serotonin receptor stimulant fenfluramine potentiates the morphine analgesia,[206] while blocking these receptors by

Table 5
EFFECTS OF DEPLETION BY α-METHYL-PARA-TYROSINE (AMT) AND PARA-CHLORO-PHENYLALANINE (PCPA) OF BRAIN CATECHOLAMINES AND SEROTONIN, RESPECTIVELY, ON THE ANALGESIA INDUCED BY OPIATES

Pretreatment	Treatment	Species	Assay	Effect	Ref.
AMT	Morphine	Rat	Tail flick	No effect	174
AMT	Morphine	Mouse	Hot plate	Antagonism	174
AMT	Morphine	Mouse	Writhing	No effect	175
AMT	Morphine	Rabbit	Tooth pulp stimulation	Antagonism	159,176
AMT	Morphine	Mouse	Hot plate	Antagonism	159,176
PCPA	Morphine	Rat	Flinch jump	Antagonism	183
PCPA	Methadone	Rat	Flinch jump	Antagonism	183
PCPA	Codeine	Rat	Flinch jump	Antagonism	183
PCPA	Propoxyphen	Rat	Flinch jump	Antagonism	183
PCPA	Meperidine	Rat	Flinch jump	No effect	183
AMT	Morphine	Mouse	Hot plate	No effect	178
AMT	Morphine	Mouse	Writhing	No effect	178
PCPA	Morphine	Mouse	Hot plate	No effect	178
PCPA	Morphine	Mouse	Writhing	Antagonism	178
AMT	Morphine	Mouse	Tail flick	No effect	177
AMT	Morphine	Mouse	Hot plate	Potentiation	161
PCPA	Morphine	Mouse	Hot plate	Antagonism	161
PCPA	Morphine	Rat	Inflamed foot	Antagonism	185
PCPA + 5-hydroxy-tryptophan	Morphine	Rat	Inflamed foot	Reversal of the antagonistic effect of PCPA	185
AMT	Morphine	Rat	Inflamed foot	Antagonism	185
PCPA	Morphine	Mouse	Hot plate	Antagonism	167
PCPA + 5-hydroxy-tryptophan + l-DOPA	Morphine	Mouse	Hot plate	Reversal of the antagonistic effect of PCPA	167
AMT	Morphine	Mouse	Hot plate	Antagonism	167
PCPA	Morphine	Mouse	Electrical foot shock	Antagonism	186
PCPA	Morphine	Mouse	Tail flick	Antagonism	184
AMT	Morphine	Rat	Vocalization after-discharge	Antagonism	75
PCPA	Morphine	Mouse	Tail flick	Antagonism	187

cyproheptadine results in attenuation of opiate analgesia.[185,188,206] Moreover, fenfluramine was found capable of inducing analgesia in itself, an effect easily reversed by cyproheptadine.[206] However, PCPA did not attenuate the morphine-induced locomotor stimulation in mice.[207] In this context it is probably worth mentioning that in rats the motor stimulation induced by dopamine injection into the n. accumbens could be antagonized by morphine, fluphenazine, and serotonin as well as this antagonism was found reversed by serotonin receptor blockers such as methysergide and cyproheptadine.[208] These findings give additional evidence that the morphine-elicited analgesic and sedative effects are closely related, i.e., probably both are mediated by serotoninergic mechanisms in either rats or mice.

For this reason it is extremely difficult to understand the role of catecholamines in opiate analgesia, especially if assuming that the sedative effect of central serotoninergic stimulation is somehow related to the antinociceptive action. As catecholamine libera-

Table 6
EFFECTS OF BRAIN MONOAMINES AND THEIR PRECURSORS ON ANALGESIA INDUCED BY OPIATES

Pretreatment	Treatment	Species	Assay	Effect	Ref.
Tryptophan sys.	Morphine	Mouse	Tail flick	No effect	173
5-hydroxytryptophan sys.	Morphine	Mouse	Tail flick	Potentiation	173
Serotonin sys.	Morphine	Mouse	Tail flick	Potentiation	173
Adrenaline sys.	Morphine	Mouse	Tail flick	Potentiation	173
5-hydroxytryptophan sys.	Morphine	Mouse	Tail pinch	Potentiation	162
l-DOPA sys.	Morphine	Mouse	Tail pinch	Potentiation	162
5-hydroxytryptophan sys.	Morphine	Mouse	Tail pinch	No effect	160
l-DOPA sys.	Morphine	Mouse	Tail pinch	No effect	160
5-hydroxytryptophan sys.	Morphine	Mouse	Tail flick	Potentiation	177
Tyrosine sys.	Morphine	Mouse	Tail flick	Potentiation	177
dl-DOPA sys.	Morphine	Mouse	Tail flick	Potentiation	177
Tyrosine sys.	Morphine	Mouse	Tail flick	Potentiation	164
5-hydroxytryptophan sys.	Morphine	Mouse	Tail flick	Potentiation	164
Serotonin i.c.v.	Morphine	Rat	Inflamed foot	Potentiation	179
l-DOPA i.c.v.	Morphine	Rat	Inflamed foot	Antagonism	179
Dopamine i.c.v.	Morphine	Rat	Inflamed foot	Antagonism	179
Noradrenaline i.c.v.	Morphine	Rat	Inflamed foot	Antagonism	179
Serotonin i.c.v.	Morphine	Mouse	Tail flick	Potentiation	190
Dopamine i.c.v.	Morphine	Mouse	Tail flick	Antagonism	190
Noradrenaline i.c.v.	Morphine	Mouse	Tail flick	Antagonism	190
Dopamine i.c.v.	Morphine	Mouse	Tail flick	Potentiation	191
5-hydroxytryptophan sys.	Morphine	Mouse	Hot plate	Potentiation	192
l-DOPA sys.	Morphine	Mouse	Hot plate	Antagonism	192
5-hydroxytryptophan sys.	Morphine	Mouse	Hot plate	Potentiation	161
l-DOPA sys.	Morphine	Mouse	Hot plate	Antagonism	161
l-DOPA sys.	Morphine	Rat	Tail flick	Antagonism	193
Dopamine i.c.v.	Morphine	Mouse	Tail flick	No effect	194
Noradrenaline i.c.v.	Morphine	Mouse	Tail flick	Antagonism	194
Adrenaline i.c.v.	Morphine	Mouse	Tail flick	Antagonism	194
Serotonin i.c.v.	Meperidine	Mouse	Tail immersion	Potentiation	189
Serotonin i.c.v.	Profadol	Mouse	Tail immersion	Potentiation	189
Noradrenaline i.c.v.	Meperidine	Mouse	Tail immersion	Antagonism	189
Noradrenaline i.c.v.	Profadol	Mouse	Tail immersion	Antagonism	189
l-DOPA sys.	Morphine	Mouse	Tail flick	Antagonism	184
5-hydroxytryptophan sys.	Morphine	Mouse	Tail flick	Potentiation	187
l-DOPA sys.	Morphine	Mouse	Tail flick	Antagonism	187

Note: i.c.v. = intracerebroventricularly; sys. = systemically (s.c., i.p.).

tion or stimulation of catecholaminergic neurones result in a powerful behavioral and autonomic excitation, i.e., in a syndrome opposite to that elicited by narcotics, dopamine and/or noradrenaline are expected to antagonize the opiate analgesia; all the more, since serotoninergic stimulation of morphine analgesia is accompanied by facilitation of the morphine-induced sedative effects.[205] Conversely, PCPA not only attenuates the antinociceptive effect of opiates (Table 5) but also induces strong motor

Table 7
EFFECTS OF ALTERATIONS IN THE CENTRAL SEROTONINERGIC AND GABAERGIC TONE ON ANALGESIA INDUCED BY OPIATES

Pretreatment	Treatment	Species	Assay	Effect	Ref.
Desipramine	Methadone	Rat	Hot plate	Potentiation	195
Aminooxyacetic acid	Morphine	Mouse	Tail compression	Potentiation	196,197
Semicarbazide	Morphine	Mouse	Tail compression	Antagonism	196,197
Bicuculline	Morphine	Mouse	Tail compression	Antagonism	196,197
Fluoxetine	Morphine	Rat	Hot plate	Potentiation	198
Doxepine	Propoxyphene	Rat	Electrical stimulation of the tail	Potentiation	199
Muscimol	Morphine	Rat	Tail flick	Antagonism	200
Amitriptyline	Morphine	Cat	Tail flick	Potentiation	201
Nortriptyline	Morphine	Cat	Tail flick	Potentiation	201
Nomifensine	Morphine	Mouse	Tail immersion	Potentiation	202
Quipazine	Morphine	Rat	Hot plate	Potentiation	203
Fluoxetine	Morphine	Rat	Hot plate	Potentiation	203
γ-Acetilenic GABA	Morphine	Mouse	Hot plate	Potentiation	204
γ-Vinyl GABA	Morphine	Mouse	Hot plate	Potentiation	204

excitation and irritability.[182] From this point of view, the data reported on the effect of oxypertine might be decisive, since this agent depletes the brain catecholamines without affecting the serotonin level. Also this agent was found to attenuate the morphine analgesia.[177,164] Thus, the narcotic analgesia cannot be ascribed exclusively to serotoninergic hyperactivity, however, i.c.v. administration of adrenaline or noradrenaline attenuate the morphine analgesia.[179,189,190,194] The direct stimulation of cental (nor)adrenergic receptors seems to inhibit the opiate analgesia. As for dopamine, it was reported by the same authors to inhibit the analgesic effect of morphne in relatively large i.c.v. doses (5 to 200 μg)[190] and to potentiate it in smaller amounts[191] in mice. It is possible that dopamine in itself potentiates the opiate analgesia, but applying it in larger doses, noradrenaline and adrenaline will be formed in sufficient amounts to inhibit the analgesia. Another tentative explanation is that giving small doses i.c.v., only those structures will be excited by dopamine which lie close to the cerebral ventricles, while giving it in higher amounts, more remotely located structures exerting opposite effects, are also stimulated. Due to their well-known incapability to pass the blood-brain barrier, neither of them may be applied peripherally, only their precursors or enzyme inhibitors which elevate or reduce their brain levels or substances, like amphetamine, which liberate the catecholamines from their presynaptic stores. Surveying these data the matter will be even more controversial.

Depletion of both catecholamines by α-methyl-p-tyrosine (AMT) inhibited the morphine analgesia in several studies,[75,159,167,174,176,185] while in others no effect or even potentiation was observed[177,175,178] (Table 5). A careful survey of these studies reveals that depletion by AMT resulted in inhibition of opiate analgesia and facilitation was found mostly in those studies[161] where the antinociceptive action was assayed by the hot plate test (Table 5). As AMT treatment in itself results in locomotor inhibition, this synergism seems to be an experimental artefact. As the effects of dopamine and noradrenaline on opiate analgesia seems to be different (Table 6), more informative are the experiments, where noradrenaline was selectively depleted by inhibition of do-

Table 8
EFFECT OF SELECTIVE DEPLETION OF BRAIN NORADRENALINE BY INHIBITION OF DOPAMINE -β- HYDROXYLASE ON MORPHINE INDUCED ANALGESIA

Pretreatment	Species	Assay	Effect	Ref.
Disulfiram	Rat	Tail pinch	Antagonism	155
Disulfiram	Rat	Tail pinch	Antagonism	154
Disulfiram	Rat	Inflamed foot	No effect	185
DDC	Rat	Tail clip	Potentiation	211
DDC	Rat	Hot plate	Potentiation	211
DDC	Mouse	Hot plate	Antagonism	192
DDC	Mouse	Hot plate	Antagonism	161
PTT	Mouse	Tail flick	Potentiation	210
PTT	Rat	Hot plate	Potentiation	209
FLA-6	Rat	Vocalization after-discharge	No effect	75

Note: DDC: diethyldithiocarbamate; PTT; 1-phenyl-3-/2-thiazolyl/-2-thiourea; FLA-63: bis-/4-methyl-1-homoperazinylthiocarbonyl/-disulfid.

pamine-beta-hydroxylase (Table 8). Selectively depleting the brain noradrenaline by 1-phenyl-3-(2-thiazolyl)-2-thiourea (PPT) generally potentiation was observed.[209,210] However, in other works inhibition was recorded if applying disulfiram as enzyme inhibitor.[154,155] The reports concerning the effect of diethyldithiocarbamate are conflicting.[161,192,211] Otherwise, Bhargawa and Way[210] found no correlation between the behavioral effects (potentiation of opiate analgesia) and enzyme inhibition upon administering PPT. Disulfiram was reported to interfere not only with the activity of dopamine-β-hydroxylase but also with the metabolism of morphine.[212] Consequently, these experiments connected with the application of enzyme inhibitors (Table 8) might result in artefacts inasmuch as the observed behavioral actions are not connected exclusively with the selective depletion of brain noradrenaline. The data summarized in Tables 6 and 8 suggest that in mice both catecholamines are involved in the opiate analgesia, while in rats dopamine plays a dominant role. In this context it is worth mentioning that pretreatment with AMT prevented the morphine-induced enhancement of performance in shuttle box[213] or the levallorphan induced disruption of operant behavior.[214] Thus, not only the antiociception but many behavioral effects of narcotics are connected with the catecholamines.

From this point of view probably those experiments deserve special attention where the neurotoxic compound 6-hydoxydopamine (6-OHDA) was used to deplete brain catecholamines. I.c.v. administered 6-OHDA had on several occasions been shown to change the antinociceptive action of morphine but the data obtained so far are very heterogeneous (Table 9). Nevertheless, as for the mouse, the data are more or less congruent, i.e., the morphine induced analgesia has been found inhibited upon 6-OHDA treatment.[215,216] This effect is generally attributed to the depletion of brain noradrenaline. Thus, in mice not the dopamine itself, but its β-hydroxylated product, i.e., noradrenaline is necessary. In rats, however, rather incongruent data have been reported (Table 9). These discrepancies might be due among others to methodological difficulties. Thus, a single treatment with 6-OHDA generally does not result in total destruction of either catecholamines. The relative degree of their depletion depends on several circumstances such as the amount of 6-OHDA administered, the age of animals used, and the contingent application of other agents, selectively blocking the neural uptake of one of the catecholamines.

Table 9
EFFECT OF DEPLETION BY 6-HYDROXYDOPAMINE[a]
OF BRAIN CATECHOLAMINES ON MORPHINE
INDUCED ANALGESIA

Species	Assay	Effect	Ref.
Rat	Hot plate	Antagonism	217
Rat	Hot plate	Potentiation	218
Rat	Tail flick	Potentiation	108
Rat	Electrical stimulation of the tail	No unequivocal effect	103
Rat	Hot plate	Antagonism	219
Rat	Tail flick	Antagonism	219
Mouse	Hot plate	Antagonism	215
Rat	Tail flick	No effect	220
Rat	Electrical stimulation of the tail	Potentiation	220
Rat	Tail compression	No effect	100
Mouse (infant)	Hot plate	Antagonism	216
Mouse (adult)	Hot plate	Antagonism	216
Rat (infant)	Tail flick	Potentiation	216
Rat (adult)	Tail flick	No effect	216

[a] 6-hydroxydopamine applied intracerebroventricularly (i.c.v.).

In experiments of Elchisak and Rosecrans[219] rats treated intracisternally at 2 weeks of age with 6-OHDA showed attenuated analgesic response some weeks later. (Under such circumstances both dopamine and noradrenaline are depleted to a certain degree.) Rats treated with 6-OHDA + desipramine showed preferential depletion of brain dopamine and a greater antagonism of morphine analgesia.[219] Thus, these authors[219] concluded, in agreement with others[216] that a critical level of brain dopamine was needed for the antinociceptive effects of opiates, while selective depletion of brain noradrenaline in adult rats did not attenuate the morphine analgesia.[216]

Injecting 6-OHDA into well-defined brain areas even the role of selected structures may be examined. Thus Nakamura et al.[108] injecting 6-OHDA into the medial hypothalamus or into the region of medial forebrain bundle observed decreased hypothalamic noradrenaline level and a potentiation of morphine analgesia. Administering 6-OHDA into the striatum, depletion of the striatal dopamine and attenuation of morphine analgesia were detected.[108] And a similar attenuation of opiate antinociception has been found upon bilateral lesioning by 6-OHDA of the substantia nigra.[165]

Probably a more physiological approach is the application of natural precursors of biogenic amines (or direct administration of the latter). As Table 6 shows, giving serotonin (i.c.v.) or its precursor 5-hydroxytryptophan, the opiate analgesia was potentiated. The lack of effect of tryptophan[173] is probably due to its limited capacity to pass the blood-brain barrier. The data showing the importance of serotonin in opiate analgesia are otherwise unanimous (Tables 3, 5, 6 and 7).

Noradrenaline — given i.c.v. — was found to inhibit the morphine analgesia in rats,[179] which is not surprising if accepting the inhibitory role of noradrenergic system in opiate analgesia in this species. However, the similar inhibition by i.c.v. administered catecholamines or morphine antinociception in mice[189,194,190] is already difficult to reconcile with the hypothetical role played by them in narcotic analgesia. L-DOPA was reported either to have no effect[160] or to potentiate[177] the analgesic effect, but the majority of authors[161,179,184,192,193] reported on inhibition. This heterogeneity of data is probably not surprising considering that the peripherally administered 1-DOPA will be converted into dopamine, noradrenaline (and adrenaline) centrally and peripherally as well and hence very different effects will be initiated simultaneously.

A pure pharmacological approach is the use of specific receptor stimulating or blocking agents, which interact with well-defined subgroups of central monoaminergic receptors. As Table 10 shows, amphetamine has been unanimously reported to potentiate the morphine analgesia. Amphetamie is generally regarded a catecholamine releasing agent but it also activates the serotoninergic mechanisms via reuptake inhibition. Thus, these data do not give information on the specific role of individual brain monoamines. Probably the same is true in relation of mescaline- and cocaine-induced potentiation of opiate analgesia (Table 10). Otherwise, the potentiation by psychostimulants of narcotic analgesia and euphoria is a well-known clinical phenomenon. This synergism is exploited by the therapeutic application of Brompton mixture containing morphine + cocaine, or by the street drug users on self-administering heroin with cocaine.

Of the more specific dopamine receptor stimulants, apomorphine has been found in our laboratory[223] to potentiate the morphine antinociception in five different assays contrary to another study.[184] While being unable to explain this discrepancy, it may be reasoned that apomorphine is expected to enhance the morphine analgesia if dopaminergic mechanisms are really involved in the action of opiates as mentioned above.

As for the dopaminergic receptor blocking agents, the data available are conflicting. The majority of authors recorded enhanced antinociception on pretreatment with haloperidol or chlorpromazine (Table 10), but some groups found the opposite effect. These latter authors regarded either the vocalization to electrical stimulation of the tail or the mechanical stimulation of the inflamed foot as pain threshold during analgesia testing[185,222] while the others applied the conventional hot plate or tail flick tests where a characteristic motor response is the indicator of pain sensation. It is tempting to speculate that during the latter procedures the well-known motor debilitating effect of neuroleptics confounds the accurate measurement of pain threshold. In a broader sense, potentiation of morphine analgesia is generally regarded as an indicator of sedative effects, whether minor tranquilizers or neuroleptics are examined. All kinds of sedatives are frequently used to potentiate the opiate analgesia in the everyday clinical practice. But of course this speculation is not necessarily unrefutable. As it has been reviewed in the previous chapter, in most species the morphine induced analgesia is only a part of a complex phenomenon, where antinociception is a manifestation of general sedation, well-being, unresponsiveness to certain internal and external stimuli. Thus, if sedation is not an unspecific "side effect" of opiates but an integral part of a behavioral syndrome then this synergism between neuroleptics and opiates also cannot be regarded as an unspecific interaction. Moreover, if chlorpromazine-like drugs do not influence the tail flick reactions if applying them alone, why should the potentiation of morphine's effect in this test be regarded as an unspecific action? It is not surprising that morphine-like drugs were for a long time classified as dopaminergic receptor blocking agents.[51] However, if morphine were a dopaminerg receptor blocking or dopamine release inhibiting agent, it should inhibit the behavioral effects of amphetamine and apomorphine and potentiate those of neuroleptics. As shown in Table 10, it is not true in the case of amphetamine if testing the analgesic effect and the data on the interaction with the classical postsynaptic dopamine receptor stimulants and inhibitors also are controversial and not only in relation of antinociception. Morphine similarly to the neuroleptics inhibits the apomorphine-induced vomiting or amphetamine + 1-DOPA induced jumping.[226] As expected of a dopamine receptor inhibitor, it enhances the chlorpromazine-induced disruption of operant behavior in rats.[227] In our recent study,[228] the conditional reflex inhibitory actions of chlorporma-

Table 10
EFFECTS OF INHIBITION OR STIMULATION OF CATECHOLAMINE AND SEROTONIN RECEPTORS ON ANALGESIA INDUCED BY OPIATES

Pretreatment	Treatment	Species	Assay	Effect	Ref.
Chlorpromazine	Morphine	Mouse	According to Hesse[225]	Potentiation	221
Chlorpromazine	Morphine	Mouse	Electrical stimulation of the tail	No effect	222
Chlorpromazine	Meperidine	Mouse	Electrical stimulation of the tail	No effect	222
Amphetamine	Morphine	Mouse	Tail flick	Potentiation	173
Mescaline	Morphine	Mouse	Tail flick	Potentiation	173
Chlorpromazine	Morphine	Mouse	Tail flick	Potentiation	172
Chlorpromazine	Morphine	Mouse	Tail flick	Potentiation	170
Chlorpromazine	Morphine	Rat	Hot plate	Potentiation	170
Perphenazine	Morphine	Mouse	Tail flick	Potentiation	170
Perphenazine	Morphine	Rat	Hot plate	Potentiation	170
α-Methyl DOPA	Morphine	Mouse	Writhing	No effect	175
Amphetamine	Morphine	Rat	Tail pinch	Potentiation	154
α-Methyl DOPA	Morphine	Rat	Tail pinch	Potentiation	154
Cocaine	Morphine	Rat	Tail pinch	Potentiation	154
Methysergide	Morphine	Mouse	Writhing	No effect	178
Methysergide	Morphine	Mouse	Hot plate	No effect	178
Phentolamine	Morphine	Mouse	Writhing	No effect	178
Phentolamine	Morphine	Mouse	Hot plate	No effect	178
Propranolol	Morphine	Mouse	Writhing	No effect	178
Propranolol	Morphine	Mouse	Hot plate	No effect	178
Amphetamine	Morphine	Mouse	Tail flick	Potentiation	177
Ephedrine	Morphine	Mouse	Tail flick	No effect	177
Propranolol	Morphine	Mouse	Tail flick	No effect	177
Amphetamine	Morphine	Mouse	Tail flick	Potentiation	164
Amphetamine	Pentazocine	Mouse	Tail flick	Potentiation	164
Amphetamine	Cyclazocine	Mouse	Tail flick	Potentiation	164
Cyproheptadine	Morphine	Rat	Inflamed foot	Antagonism	185
Chlorpromazine	Morphine	Rat	Inflamed foot	Antagonism	185
Haloperidol	Morphine	Rat	Inflamed foot	Antagonism	185
Phenoxybenzamine	Morphine	Rat	Hot plate	Potentiation	209
Phentolamine	Morphine	Rat	Hot plate	Potentiation	209
Propranolol	Morphine	Rat	Hot plate	No effect	209
Practolol	Morphine	Rat	Hot plate	No effect	209
Clonidine	Morphine	Mouse	Tail flick	Potentiation	224
Phenoxybenzamine	Morphine	Mouse	Tail flick	No effect	184
Propanolol	Morphine	Mouse	Tail flick	No effect	184
Haloperidol	Morphine	Mouse	Tail flick	Potentiation	184
Pimozide	Morphine	Mouse	Tail flick	Potentiation	184
Apomorphine	Morphine	Mouse	Tail flick	Antagonism	184
Pimozide	Morphine	Rat	Vocalization afterdischarge	Antagonism	75
Chlorpromazine	Morphine	Rat	Vocalization afterdischarge	Antagonism	75
Phenoxybenzamine	Morphine	Rat	Vocalization afterdischarge	Potentiation	75
Yohimbine	Morphine	Rat	Vocalization afterdischarge	Potentiation	75
Phenoxybenzamine	Morphine	Mouse	Tail flick	No effect	187
Propanolol	Morphine	Mouse	Tail flick	No effect	187
Apomorphine	Morphine	Mouse	Tail flick	Antagonism	187
Haloperidol	Morphine	Mouse	Tail flick	Potentiation	187
Pimozide	Morphine	Mouse	Tail flick	Potentiation	187

Table 10 (continued)
EFFECTS OF INHIBITION OR STIMULATION OF CATECHOLAMINE AND SEROTONIN RECEPTORS ON ANALGESIA INDUCED BY OPIATES

Pretreatment	Treatment	Species	Assay	Effect	Ref.
Apomorphine	Morphine	Rat	Tail flick	Potentiation	223
Apomorphine	Morphine	Rat	Inflamed foot	Potentiation	223
Apomorphine	Morphine	Mouse	Hot plate	Potentiation	223
Apomorphine	Morphine	Mouse	Writhing	Potentiation	223

zine and morphine were found simply additive and the detailed statistical analysis failed to detect either synergism or antagonism. Thus it was concluded that the disrupting effects of morphine and chlorpromazine (or that of haloperidol) were mediated via events independent of each other.[228] As for the spontaneous locomotion and stereotypy, the data available are conflicting; morphine has been reported both to inhibit and potentiate the effect of apomorphine.[153,226,229] In cats, the charactristic behavioral excitation observable upon morphine treatment may be prevented by CNS catecholamine depletors (reserpine and tetrabenazine) or dopamine receptor blocking agents (haloperidol and chlorpromazine).[129,230] These observations indicate that in this species morphine releases dopamine, which in turn excites the central dopaminergic receptors, responsible for the feline mania.

Returning to the alleged role of adrenergic receptors in morphine antinociception, the usage of adrenergic receptor blockers seems to be a logical way. Data summarized in Table 10 suggest that β-adrenergic receptors play no significant role in morphine analgesia. Opiate analgesia was not found in either study to be modified upon pretreatment with propranolol.[177,178,184,187,209] However, propranolol was reported to enhance the acute toxicity of morphine,[231] showing that probably different mechanisms mediate the pharmacological and toxic effects of opiates. As for the α-adrenerg blockers, they have been reported either to potentiate[75,209] or to have no effect[178,184,187] on opiate analgesia. Considering the inhibitory effect of centrally administered (nor)adrenaline on opiate analgesia (Table 6) in rats, in this species potentiation is expected. The reported ineffectiveness of phenoxybenzamine and phentolamine is difficult to interpret (Table 10); all the more, since these substances were reported to have naloxone reversible analgesic action due to a certain affinity to the opiate receptors.[232-234] (Unless they are supposed to behave as partial agonist-antagonists in relation of pure narcotics, having analgesic action if given alone but attenuating the effect of stronger agonists.)

B. Cholinolytics and Cholinomimetics

Opiate analgesia also may be modified by pharmaca exerting their actions independently of the monoaminergic neurones. First of all drugs modifying the central cholinergic tone deserve special attention. The first observations that cholinomimetic agents produce behavioral effects, especially analgesia, were made in humans.[235] As the Table 11 shows scentral muscarinergic receptor agonists and cholinesterase inhibitors potentiate the opiate analgesia in a variety of tests involving several species. On the other hand atropine has been repeatedly reported to have no effect or rather to inhibit the antinociceptive effect of morphine (Table 11). Thus, a very simple conclusion arises, increment in central cholinergic tone is tantamount to potentiation of central opiate receptor mediated mechanism. However, a more thorough survey of literature does not confirm such a simplified conclusion. An increase in central and peripheral cholinergic tone cannot be the neurochemical mechanism underlying the antinociceptive action of opiates, since they inhibit the acetylcholine release in the specifically opiate

Table 11
EFFECTS OF STIMULATION OR INHIBITION OF CHOLINERGIC TRANSMISSION ON ANALGESIA INDUCED BY OPIATES

Pretreatment	Treatment	Species	Assay	Effect	Ref.
Neostigmine	Morphine	Rat	Tail clip	Potentiation	236
Scopolamine	Morphine	Mouse	Tail pinch	No effect	72,237
Scopolamine	Meperidine	Mouse	Tail pinch	No effect	72,237
Psysostigmine	Morphine	Mouse	Tail flick	Potentiation	238
Physostigmine	Morphine	Mouse	Tail flick	Potentiation	164
Physostigmine	Pentazocine	Mouse	Tail flick	Potentiation	164
Physostigmine	Cyclazocine	Mouse	Tail flick	Potentiation	164
Physostigmine	Morphine	Mouse	Writhing	No effect	177
Oxotremorine	Morphine	Mouse	Writhing	No effect	177
Physostigmine	Morphine	Mouse	Writhing	Potentiation	235
Oxotremorine	Morphine	Mouse	Electrical stimulation of the tail	No effect	235
Atropine	Morphine	Mouse	Electrical stimulation of the tail	No effect	235
Physostigmine	Morphine	Mouse	Hot plate	Summation	192
Oxotremorine	Morphine	Mouse	Hot plate	Summation	192
Physostigmine	Morphine	Mouse	Tail flick	Potentiation	190
Physostigmine	Nalorphine	Mouse	Tail flick	Potentiation	190
Physostigmine	Pentazocine	Mouse	Tail flick	Potentiation	190
Acetylcholine	Morphine	Mouse	Tail flick	Potentiation	194
Physostigmine	Morphine	Mouse	Tail flick	Potentiation	194
Scopolamine	Morphine	Monkey	Shock titration	Potentiation	77
Physostigmine	Morphine	Monkey	Shock titration	Potentiation	77
Arecoline	Morphine	Monkey	Shock titration	No effect	77
Atropine	Morphine	Mouse	Tail flick	Antagonism	184
Physostigmine	Morphine	Mouse	Tail flick	Potentiation	184
Atropine	Morphine	Rat	Vocalization afterdischarge	No effect	75
Atropine	Morphine	Mouse	Tail flick	Antagonism	187
Physostigmine	Morphine	Mouse	Tail flick	Potentiation	187
Atropine	Morphine	Rat	Electric foot shock	No effect	239
Pilocarpine	Morphine	Rat	Electric foot shock	Summation	239

sensitive isolated organ preparations (Chapter 1, Volume I) and in brain slice preparations (Chapter 1, Volume II) as well. It is well known that physostigmine at higher dose levels depresses the general motor behavior. Therefore, it is difficult to separate the alleged antinociceptive effect from the general motor debilitating action in tests, where a motor reaction signals the pain threshold. They actually produce catalepsy and also depress both the mono- and polysynaptic spinal reflexes[77] and these effects may easily confound the results of analgesia testing. Interestingly enough, pilocarpine and arecoline were found to antagonize the morphine-induced catalepsy.[239] Moreover, arecoline did not potentiate the analgesic effect of morphine in the shock titration test in monkeys and physostigmine did it only in doses causing general motor depression.[77] Otherwise, almost all the experimental studies concerning the interaction of muscarinergic agents with opiate analgesia were made in rodents (Table 11). And in rodents, contrary to primates, the cholinomimetics exert analgesic action of their own. Thus, the observed "potentiation" may simply be due to the addition of two independent events; all the more, since the analgesic action of cholinomimetics cannot be antagonized by naloxone.[235]

To make the matter even more controversial, some antimuscarinergic drugs were reported to potentiate the opiate analgesia. As for atropine, inhibition or no effect were reported, while in the case of scopolamine mostly potentiation was found (Table 11). In the clinical practice atropine or scopolamine are frequently coadministered with opiates in order to diminish the latters' unpleasant autonomic side effects. They probably would not be so frequently administered simultaneously if the combination were really deliterious in relation to opiate analgesia.

Considering the complex behavioral effects of atropine-like drugs, these contradictions are not very surprising. Probably Parke's studies[72,237] may help to resolve these contradictions at least in part. He did not find the morphine analgesia modified by scopolamine in the classical tail flick test.[237] However in the Haffner test (i.e., pinching the mouse tail with artery forceps) scopolamine potentiated the effect of morphine as far as the squeak response was concerned, but did not alter the motor reaction. Supposing the motor reactions in the tail flick and tail pinch procedures or upon electrical stimulation of the tail are primarily spinal reflexes, while vocalization is a supraspinal response,[56,57,74] a very simple explanation arises according to cholinolytics modify the antinoceptive reactions at supraspinal level. Thus antinociception may be achieved by a number of different pharmacological manipulations, each of them effecting distinct neurochemical and neurophysiological processes unique to a given receptor population, structure, or level of interaction.

C. Other Drugs

Disregarding several studies using the nonspecific phenylquinone writhing test neither pentobarbitone[240] nor the benzodiazepines[241] were found to modify the morphine analgesia at specific dose levels. Otherwise, the minor tranquilizers at high, sedative dose levels "potentiate" the opiate analgesia similarly to other classes of CNS depressants. This effect is, however, regarded as a nonspecific manifestation of sedative action; surely enough, potentiation of morphine analgesia is a frequently used test to measure the sedative action of CNS drugs.

It is already well established that antidepressants potentiate the analgesic effect of narcotic drugs[198,199,201-203,242] (Table 7) and they have a certain antinociceptive effect of their own.[243] Potentiation by imipramine of propoxyphene-induced analgesia has been observed in our laboratory too (unpublished observations[244]). The mode of antinociceptive action of antidepressants is nowadays intensively investigated, an issue which lies outside the scope of this monograph. But let us mention here that there are several possible explanations for this effect. Being strong inhibitors of serotonin (and noradrenaline) uptake, antidepressants might act via potentiation of the serotoninergic mechanisms. Furthermore, in several experiments the facilitation of adrenergic (noradrenergic) mechanisms, or their sedative effects might also contribute to the potentiation of opiate analgesia.

It is more difficult to interpret the opposite action, namely the well-known clinical antidepressant effect of opiates (Chapter 10, Volume III). Opiates are active in several pharmacological tests regarded specific for tricyclic antidepressants such as reversal of reserpine-induced hypothermia, ptosis, and behavioral depression. These effects are of special interest in view of the alleged antidepressant action of endogenous opioids (Chapter 3, Volume II). Nevertheless, it must be taken into consideration that the above discussed behavioral stimulant, sympathomimetic and stressor effects of opiates might easily explain these quasi antidepressant actions.

In some studies GABA has been implicated in opiate analgesia.[196,204,245] Namely increasing the brain level of GABA by aminooxyacetic acid resulted in potentiation of morphine analgesia, while decreasing its activity by semicarbazide led to its attenuation.[196,197] Bicuculline, the GABA-erg antagonist also inhibited the morphine antinociception.[196] The GABA agonist muscimol was reported to potentiate the morphine analgesia.[200]

Otherwise, naloxone was reported to potentiate the bicuculline-induced convulsions and to attenuate the facilitation of unit activity induced by GABA in nucleus accumbens and tuberculum olfactorium.[245] Thus, the morphine-induced behavioral depression seems to be related to a certain degree to the increased GABA-ergic activity.

The contingent role of prostaglandins has not been extensively studied yet. Actually PGE_1 was reported both to potentiate[246] and to attenuate[247] the analgesic action of morphine.

VII. LOCALIZED INTRACEREBRAL APPLICATION

Even a superficial survey of data on the effects of locally applied intracerebral opiate injections may give explanation for the contradictory reports concerning the behavioral effects elicited by systemic application of narcotic drugs. Considering that activation by morphine of different brain structures results in very different behavioral and autonomic effects, it will be easy to understand that all conclusions drawn from the experimental data obtained by systemic application of opiates should be revised or at least regarded as the summation of many individual and more or less independent events.

The first systematic study on the neuroanatomical focus of morphine action was carried out by Herz and his co-workers.[248] Placing a plug into the aqueduct, the spread of i.c.v. injected drug solution was restricted and the effect of opiates administered into well-defined segments of the liquor space could be studied. And it turned out that no analgesia might be elicited through the lateral or third ventricle, i.e., the forebrain structures were not involved in opiate analgesia[248] (Table 12). And only weak analgesia was found if injecting opiates into the cisterna magna without spread into the fourth ventricle and aqueduct, suggesting that cerebellar, lower medullary and spinal structures alone were not capable to bring about strong antinociception.[248] The critical importance of periaqueductal and posterior periventricular area in the floor of the fourth ventricle has been confirmed by Jacquet and co-workers[123,128] Pert and Yaksh,[249] and by others (Table 12). Yaksh et al.[256] demonstrated that injecting morphine into the different regions within the PAG even the analgesic effect was more or less localized over the surface of the body.

Otherwise, the antagonizability by naloxone has been demonstrated in almost all laboratories examining the action and one group[256] reported also on tolerance development upon repeated application of morphine into the PAG. Thus, the effects meet all the main criteria of opiate specific actions. Moreover, the exceptional importance of PAG and raphe nuclei (see below) is in good agreement with the results of experiments concerning the effect of selective lesioning of different brain structures (Chapter 2, Volume I).

Consequently, according to the data obtained upon localized intracerebral microinjection of opiates, their main point of attack seems to be the PAG,[123,128,248,249,252,256] the raphe nuclei,[91,265,266] posterior hypothalamus,[248,250,251] and the n. reticularis (para)gigantocellularis of the medulla oblongata.[261,267] In monkeys beside these structures centered just around the midline, several other regions are also highly sensitive to opiates such as the lateral reticular nucleus of the brain stem, substantia nigra, centromedian, and parafascicular nuclei of the thalamus.[249]

As for the posterior hypothalamus, local application of opiates into this region resulted in variably in analgesia (Table 12), though not very much has been reported on the accompanying behavioral symptoms. These findings are not surprising considering that the hypothalamus contains opiate receptors in high concentration (Chapter 1, Volume III), and this area is a frequently used site of self-stimulation (Chapter 8, Volume III).

The PAG is another area where the presence of opiate receptors is well demonstrated (Chapter 1, Volume III) and known to contain self-stimulation reinforcing loci. Local

Table 12
ACUTE BEHAVIORAL EFFECTS UPON LOCALIZED INTRACEREBRAL APPLICATION OF OPIATES

Drug	Site of application	Species	Observed effect	Ref.
Morphine	Hypothalamus	Rat	Analgesia, catalepsy, respiratory depression, exophtalmos, hypothermia	250
Morphine	Mamillary nuclei	Rat	Hyperactivity, hyperthermia	250
Morphine	Hypothalamus (periventricular gray matter)	Rat	Analgesia	251
Morphine or fentanyl	i.c.v. but restricted to the lateral or third ventricle by a plug inserted into the aqueduct	Rabbit	No antinociception, catalepsy-like state alternating with attentative behavior (raising the head, sniffing, pricking the ears)	248
Morphine or fentanyl	Aqueduct or fourth ventricle	Rabbit	Strong analgesia, marked behavioral and respiratory depression	248
Morphine or fentanyl	Intracisternally without spread into the fourth ventricle and aqueduct	Rabbit	Slight analgesia, weak behavioral and respiratory depression	248
Morphine or fentanyl	Septum, comissura fornicis, dorsal hippocampus, thalamus pars dorsalis and lateralis	Rabbit	No analgesia	248
Morphine	Mesencephalon, subthalamus	Rabbit	Analgesia	248
Morphine	Hypothalamus pars medialis	Rabbit	Strong analgesia, sedation	248
Morphine	Hypothalamus pars posterolateralis	Rabbit	Strong analgesia, increased vigilance	248
Morphine	PAG	Rabbit	Strong analgesia, motor excitation, hyperexcitability, vocalization	248
Morphine	Posterior hypothalamus, third ventricle	Rat	Analgesia	128
Morphine	Medial septal nucleus, candate nucleus	Rat	Hyperalgesia, hyperexcitability, violent jumps to auditory and visual stimuli	128
Morphine	PAG	Rat	Hyperreactivity to auditory, visual, and tactile stimuli, no reaction to nociceptive stimuli	123
Levorphanol or methadone or etorphine	PAG	Rat	Analgesia, but no hyperreactivity to non-noxious stimuli	123
Morphine	PAG	Rat	Analgesia and hyperreactivity	252
Morphine	Globus pallidus	Rat	Weak gnawing response after long latency	253
Morphine	Ventral thalamus	Rat	Strong excitation, gnawing, biting, running, turning, sometimes convulsions after long latency	253

Table 12 (continued)
ACUTE BEHAVIORAL EFFECTS UPON LOCALIZED INTRACEREBRAL APPLICATION OF OPIATES

Drug	Site of application	Species	Observed effect	Ref.
Morphine	Caudate nucleus and putamen	Rat	No effect	253
Methadone	Ventral thalamus	Rat	Effects as seen with morphine	253
Meperidine	Ventral thalamus	Rat	No effect	253
Morphine	PAG and periventricular gray matter	Monkey	Strong analgesia	249
Morphine	Thalamus, intralaminar nuclei	Monkey	Strong analgesia	249
Morphine	n. subthalamicus	Monkey	Strong analgesia	249
Morphine	Substantia nigra	Monkey	Moderate analgesia	249
Morphine	Lateral reticular nucleus of the brain stem	Monkey	Moderate analgesia	249
Morphine	Parafascicular nucleus of the thalamus	Monkey	Moderate analgesia	249
Morphine	Mesencephalic reticular formation	Rat	Ipsilateral rotation, no analgesia	254, 255
Morphine	PAG	Rat	Analgesia	256
Morphine	Anterior and medial thalamus	Rat	No antinociception	256
Morphine	Mesencephalon (tectum and tegmentum)	Rat	No antinociception	256
Morphine	Septum	Rat	No antinociception	256
Morphine	PAG	Rat	Analgesia	257
Morphine	PAG	Rat	Strong analgesia	258
Morphine	Lemniscus medialis	Rat	Strong analgesia	258
Morphine	n. mediodorsalis thalami	Rat	Strong analgesia	258
Morphine	Brainstem reticular formation	Rat	Moderate antinociception	258
Morphine	n. lateralis septi	Rat	Moderate antinociception	258
Morphine	n. mamillaris lateralis	Rat	Moderate antinociception	258
Morphine	Colliculus sup. and inf.	Rat	Moderate antinociception	258
Morphine	n. accumbens	Rat	Hypermotility 3-6 hr after treatment	121
Morphine	n. accumbens	Rat	Analgesia, catalepsy	259
Morphine	Caudate nucleus	Rat	No effect	259
Morphine	Globus pallidus	Rat	Weak catalepsy	259
Morphine	Substantia nigra unilaterally	Rat	Contralateral circling	157, 158
Morphine	Substantia nigra bilaterally	Rat	Stereotypy (gnawing, sniffing, biting)	157, 158
Morphine	Caudate nucleus	Rat	No effect	157, 158
Morphine	Hypothalamus	Rat	Inhibition of drinking	260
Morphine	Intrathecally	Rat	Analgesia and hyperthermia, but no behavioral or respiratory depression	12
Morphine	n. reticularis gigantocellularis	Rat	Analgesia but no behavioral depression	261
Morphine	n. accumbens	Rat	Catalepsy	262
Morphine	Striatum	Rat	No effect	262
Morphine	n. accumbens	Rat	Initial sedation followed by hypermotility	263

Table 12 (continued)
ACUTE BEHAVIORAL EFFECTS UPON LOCALIZED INTRACEREBRAL
APPLICATION OF OPIATES

Drug	Site of application	Species	Observed effect	Ref.
Morphine	n. septi medialis	Rat	No effect	264
Morphine	n. raphe magnus	Rat	Analgesia without behavioral depression	265,266
Morphine	n. reticularis paragigantocellularis	Rat	Analgesia without behavioral depression	267
Morphine	Caudate nucleus	Rat	Moderate analgesia	268
Levorphanol	Caudate nucleus	Rat	Moderate analgesia	268
Meperidine	Caudate nucleus	Rat	Moderate analgesia	268

application of opiates into this area elicits not only analgesia but also catalepsy, strong behavioral and respiratory depression as well (Table 12), thus almost all behavioral effects regarded specific for opiates. But interestingly not only depressant effects may be elicited from this area. First Herz et al.[248] mentioned, then Jacquet and Lajtha[123,128] analyzed in detail a perplexing phenomenon: the irresponsiveness to any kind of painful stimulus was accompanied by an unawaited hyperreactivity (vigorous jumping) to auditory, visual, and tactile stimulation. This hyperreactivity, called latter "explosive motor behavior"[124] could be reversed by locally or systemically applied naloxone according to their first study,[128] a statement revoked in a subsequent report.[269] (Of course the depressant effects were found naloxone sensitive in all their studies mentioned above.) Upon peripheral administration hyperactivity frequently follows the morphine-induced CNS depression (Chapter 2, Volume I). However, administering it directly into the PAG the sedative and stimulant actions seem to appear simultaneously according to these observations (see above). On the other hand, the same authors emphasized that within the mesencephalon, i.e., in the PAG and in the adjoining reticular formation there were sites from where both analgesia and hyperexcitability, or only analgesia or the excitatory response could be elicited.[123,128,254,269]

Upon unilateral morphine application in several cases, especially if morphine was not given into the PAG but into the neighboring reticular formation, unilateral turning was observed.[254,269] This morphine-induced turning behavior could not be modified with either peripherally given pimozide, phentolamine or propranolol, or locally administered dopamine.[254] However, locally applied atropine or noradrenaline inhibited while carbachol potentiated this rotational movement.[254] Consequently, the morphine-induced hyperexcitability syndrome seems to be a phenomenon initiated by cholinergic mechanisms. This excitation could also be elicited by high doses of (+)-morphine, and naloxone did not reverse the syndrome, and even more interestingly the syndrome also could be evoked by ACTH (1-24).[124] Thus the authors concluded on the existence of separate endorphin and ACTH receptors, the former being naloxone sensitive and responsible for analgesia, the others eliciting hyperactivity upon being activated by its natural ligands (ACTH) or by any enantiomers of morphine.[124,269] Surprisingly, other opiates such as levorphanol, methadone, or etorphine failed to elicit the explosive motor behavior, a finding which was explained by the rapid diffusion of the morphine congeners from the site of application due to their high lipophilicity.[123,128] Beside the basic finding (i.e., simultaneous occurrence of depressant and excitatory symptoms), which has observed also by others,[248,270] many details of these findings and especially the hypothesis built on them requires further clarification and confirmation.

Of the other structures connected with the PAG or brain stem reticular formation, the raphe nuclei deserve special attention. Microinjection of morphine into the raphe complex was repeatedly reported to induce analgesia.[265,266,271]

Interestingly, this analgesic effect was not accompanied by either CNS depressant or excitatory behavioral signs and no electroencephalographic alterations were found.[265] Thus the analgesia elicited by chemical stimulation of the raphe complex is not connected with the characteristic behavioral symptoms observable upon systemic or more cranial intracerebral application of opiates. Actually, raphe is one of the few brain structures from where analgesia may be elicited without any concomitant gross behavioral alteration. Moreover, this analgesic effect of morphine given into the raphe nuclei seems to be dependent on serotoninergic mechanisms. Namely, its antinociceptive effect was abolished or at least attenuated by serotonin receptor blocking agents among others by i.p. administered cinanserine[265] or intrathecally given methysergide.[272] The analgesic action could also be antagonized by naloxone.[265,266] Microinjection of morphine into the PAG evoked serotonin release from the spinal cord.[271] And also direct intrathecal morphine administration resulted in analgesia,[12,273] while intrathecal naloxone injection only partially reversed the opiate analgesia.[274] Thus the analgesia initiated by chemical stimulation of the PAG or the raphe nuclei seems to be mediated by activation of descending serotoninergic pathways. A very attractive hypothesis, which explains many experimental findings, among others the potentiation or attenuation of morphine analgesia by serotonin receptor stimulation or inhibition, respectively (Chapter 2, Volume I), and the diminution of morphine's analgesic potency in spinal animals as well (Chapter 2, Volume I).

Of course not all the descending pathways activated by morphine are serotoninergic. Thus Yaksh and Tyce[271] reported that the analgesia induced by morphine microinjected into the PAG also might be inhibited by intrathecally given phentolamine and potentiated by adrenergic agonists. (Considering the functional connections between PAG and the raphe nuclei,[112] the two structures may probably be discussed as a functional entity.)

Beside PAG and the raphe nuclei a third area, claimed to elicit strong analgesia upon local stimulation, lies in the medulla. This area is located in the nucleus reticularis gigantocellularis[261] and nucleus reticularis paragigantocellularis[267] as shown by a Japanese group. Considering the amount of morphine needed to elicit analgesia via this area (less than 1 μg),[261,267] this region seems to be the most sensitive one to opiates. Otherwise, stimulation of these nuclei induces analgesia also without behavioral symptoms.[261,267] Thus chemical activation of opiate receptors caudally to the PAG results in analgesia without any other motor effect. Finally let us mention also here that the PAG, the raphe complex, and nucleus reticularis (para) gigantocellularis of the medulla are those regions from where analgesia can be elicited also by focal electrical stimulation[267,275] (Chapter 8, Volume III).

In the case of intrathecal administration probably higher doses of morphine or fentanyl are needed to induce analgesia, but otherwise the time-course and the slope of dose-response curves are as usual.[12,273] The relatively higher doses needed prove that in intact animals the supraspinal mechanisms contribute to the analgesic actions of opiates. Otherwise, the intrathecally induced antinociception is the "purest", so far not only the customary behavioral depression or excitation are missing but also the majority of autonomic effects, such as respiratory depression, bradycardia, and hypotension.[12] This is not surprising since opiates administered into the spinal cord obviously did not gain access to the opiate sensitive respiratory and cardiovascular centers located in the medulla or even more cranially. Therefore morphine is more and more frequently administered intrathecally in the clinical practice.[276]

The alleged participation of diencephalic structures in mediation of opiate analgesia is a rather controversial issue. Even more so in the cases of striatum and substantia nigra both rich in opiate receptors (Chapter 1, Volume III). Injection of morphine into the rostral part of the ventricular system[248] or its direct application into the striatum[128] was reported to induce increased attention, hyperexcitability but no analgesia. In other studies weak catalepsy or no effect was observed upon injecting it into the nucleus caudatus or globus pallidus.[253,259,262] In a recent study, however, intracaudate injection of morphine resulted in moderate analgesia, which could be antagonized by both naloxone and apomorphine.[268] In this context it is worth recalling that also Jacquet and Lajtha[123,128] observed an elevation of the pain threshold if morphine was given intracaudally in the dose of 1 µg, while hyperalgesia was elicited by them upon raising the dose to 10 µg. Otherwise, in the above-mentioned experiments of Jurna and Heinz[268] not only morphine and levorphanol but also the latter's inactive enantiomer (dextrorphan) elicited the lengthening of tail flick latency time. In this experiment also haloperidol had a certain "antinociceptive" effect.[268] Thus, impairment of the dopaminergic functions of the striatum may result in a moderate inhibition of antinociceptive reactions; an effect, which may explain the potentiation of opiate analgesia by neuroleptics (Chapter 2, Volume I) but hardly the analgesic effect of the opiates themselves. Naloxone given into the caudate nucleus also diminished to a certain degree but did not abolish the analgesic effect of intraperitoneally administered morphine.[268]

No analgesia was detected when microinjecting morphine into the substantia nigra, but contralateral circling behavior or stereotypy upon bilateral application, similar to the dopamine agonists was seen.[157,158] Since the effects are naloxone sensitive, they are obviously initiated by the opiate receptors known to be present in this nucleus (Chapter 1, Volume III).

Thus opiates seem to inhibit the dopaminergic transmission if giving them directly into the striatum and facilitate it via nigral mechanisms, but none of these actions seems to be related to their cataleptogenic effect. The latter may be elicited from the PAG (as mentioned above) and via the nucleus accumbens. When injecting morphine into the latter, analgesia and catalepsy develop, which can be reversed by naloxone or naltrexone.[259,262,263,277] This period of analgesia and behavioral depression is always followed by long-lasting hyperactivity.[121,263,277] Otherwise, hypo- and hypermotility may be elicited through the nucleus accumbens also by dopamine antagonists and agonists, respectively.[259] Nucleus accumbens is probably the only brain structure from where both phases of opiate-induced motor changes can be elicited just as upon systemic injection. A further conclusion arising from these observations[121,259,263,277] is that these much debated behavioral effects of opiates are either independent of the striatal mechanisms, or the latters are only secondarily involved. According to Costall and co-workers[263] the morphine-induced catalepsy (elicited via nucleus accumbens) can be reversed not only by naloxone but also by peripherally administered cyproheptadine. Thus, opiate-induced catalepsy has serotoinergic mechanism even if it is initiated via this mesolimbic structure. In this context it is interesting that lesioning of the medial and dorsal raphe nuclei prevented the morphine-induced catalepsy just as lesions placed into the central nucleus of the amygdala.[150,278] It is tempting to speculate that injecting opiates into the accumbens the raphe nuclei will finally be activated. Otherwise, the locomotor excitation succeeding the phase of catalepsy can be reversed by alpha adrenergic receptor blockers (piperoxane, aceperon), dopamine antagonists (fluphenazine), and also by α-methyl-p-tyrosine,[277] i.e., by inhibition of catecholaminergic neurones. Thus the first phase of morphine action induced via nucleus accumbens reflects a predominance of serotoninergic functions, while during the second phase catecholaminergic mechanisms prevail. This is actually the final conclusion obtained by others in a different way. Bergman et al.[156,253] observed that administering opiates or apomorphine into the ventral nucleus of the thalamus resulted in general excitation and gnawing, but these effects appeared with a latency of 1 to 2 hr. These effects could be antagonized not only by dopamine receptor blockers (chlorpromazine, haloperidol, pimozide) but also by peripherally administered morphine, and potentiation was observed upon previous depletion of brain serotonin by p-chloro-phenylalanine.[156,253] They concluded on simultaneous activation of serotoninergic and catecholaminergic mechanisms upon opiate treatment.[253]

The implications of these experiments are a bit broader. Perhaps opiates simultaneously activate certain catecholaminergic and serotoninergic mechanisms which are mutually antagonistic. Accordingly, the consequences of dopaminergic (adrenergic) stimulation appear only after dissipation of the first phase. And of course only the first phase of serotoninergic prevalence is accompanied by CNS depression and catalepsy (and analgesia), but both sets of events are initiated by opiate receptors. Accordingly, the long latency of morphine-induced gnawing were due to the preponderance of the initial serotoninergic mechanisms. Both the suppression of dopaminerg mechanisms in the first phase and their activation in the second phase seem to be independent of the striatal mechanisms, hence the inability of atropine to antagonize the cataleptogenic effect of opiates[146] and the importance of extrastriatal structures such as n. accumbens, ventral thalamus, and PAG in elicitation of "typical" behavioral effects of opiates. Since opiates elicit dopaminergic stimulation via ventral thalamus[156,253] an effect antagonistic to analgesia, it is easy to understand why the antinociceptive action of opiates is facilitated upon destruction of this area.[110] Since opiates administered into the striatum are ineffective (Table 12) and its lesioning does not modify significantly themorphine analgesia,[110] this structure can hardly play an important role in the antinociceptive effect of opiates. Otherwise the dopamine receptor blockade results in potentiation of narcotic analgesia (Chapter 2, Volume I). This might reflect inhibition of opiate analgesia by dopamine-activated pathways, or in the light of the functional antagonism between serotoninergic and dopaminergic cells it might be the consequence of disinhibition of the serotoninergic neurones contributing to opiate analgesia. Since dopamine agonists were also found to potentiate the narcotic analgesia (Chapter 2, Volume I), neither explanation can be accepted.

Thus, administering opiates either into the n. accumbens, ventral thalamus, striatum, or PAG, different behavioral events corresponding to different neurotransmitter systems and possibly to different neurone populations will be simultaneously activated. These findings contradict the proposal of several behavioral pharmacologists who regard the opiate analgesia as only a secondary event, a manifestation of general CNS depression, areflexia, catatonia, etc. These data show that the individual components of opiate-induced behavioral syndrome are mediated by different brain structures in spite of their simultaneous occurrence. This conclusion might be self-evident for the neuropharmacologists who are aware of the differential brain organization of different reflexes or for the psychopharmacologists, who know that many substances induce catalepsy, rigidity, exophthalmos without analgesia. The separability of the dramatic increment in pain threshold from other opiate-induced effects is, however, surely not so conspicuous for the clinicians or for research workers prone to speculate rather in psychological terms.

According to the common clinical practice opiates do not abolish the pain sensation, only the concomitant negative emotions. The cancer patients clearly perceive such relatively moderate pain as a pinprick, but they experience no suffering due to their sickness if treated by opiates.

Also the relative irresponsiveness of certain schizophrenic patients to painful stimuli is obviously of emotional origin.

Under several emotionally stressful conditions, e.g., wounded soldiers on the battlefield, women during labor, even healthy persons do not experience pain or at least the pain sensation is blunted.

In humans, disregarding the effects of some partial agonists, there is no opiate-induced analgesia without concomitant euphoria, a word referring unequivocally to the characteristic psychological changes induced by morphine derivatives. Thus, for psychologists, psychiatrists, and clinicians, opiate-induced analgesia appears as a manifestation of a specific psychic state.

The state dependent learning, the capability of narcotic drugs to serve as discriminative stimuli prove that they induce a peculiar subjective state also in experimental animals and no experimental data published yet according to the stimulus property of narcotic drugs could have been confounded without the concomitant loss of analgesic potency.

Nevertheless, there are some experimental data proving that the analgesic action of opiates cannot be interpreted in purely behavioral or psychological terms.

First, if opiates do not influence the primary pain mechanisms, only the concomitant negative emotions, they should be ineffective in spinal or in caudally decerebrated animals, since the emotional reactions are organized obviously at higher levels. As discussed above, the analgesic effect of opiates can clearly be detected in spinal animals too. Moreover, the primary sites of opiate actions are scattered in the brain stem, i.e., caudally from the forebrain structures organizing the emotional reactions in mammalian species.

Secondly, even almost four decades after its introduction, the rat tail flick test is probably still the most reliable assay to measure the opiate-specific action in vivo. Nevertheless, this tail withdrawal reaction is an entirely spinal reflex, which is not only maintained in spinal preparation, but hardly influenced by spinal transection.

Through the discovery of opiate receptors and by elucidating their wide distribution in the different brain structures, these contradictions will probably be easier to resolve. It is conceivable that opiates simultaneously inhibit the pain reflexes directly at spinal level, indirectly via the descending spinopetal pathways, and suppress the accompanying negative emotions through the corresponding limbic structures so rich in opiate receptors. The clinicians and the behavioral pharmacologist see the latter mechanisms but the in vivo analgesia testing is mainly based on reactions realized by lower brain structures.

VIII. EFFECTS OF OPIATES ON CONDITIONED BEHAVIOR

Modification of learning or memory does not belong to the most specific effects of opiates. Nevertheless, since opioid peptides seem to be involved in these processes (Chapter 3, Volume II), the effects of opiates on conditional reactions deserve a short survey. It must be emphasized that the conditioned behavior can be influenced in many ways. Not only the alteration of the memory itself but any change in the level of vigilance, in emotionality or in sensory functions can influence the results of these experiments and the "aspecific" actions may completely confound the direct effect exerted on memory. Since in animal experiments mostly motor reactions are conditioned, any change in motor functions and the spontaneous locomotion may also alter the results. Consequently, it is not surprising that depending on the experimental paradigm used, very different data have been published.

In humans, codeine facilitated the acquisition in a paired-association learning task.[279] Facilitation has been observed also in animal studies where one-trial appetitive or aversive learning was examined, administering morphine after the training session and recording the effect in the test session.[280,281] The observed facilitation of acquisition and recall has been interpreted in terms of a possible involvement of opioids in reward mediation.[279,281] There are, however, some conflicting reports too. Kesner[282] observed no effect on retention of avoidance conditioning if opiates were administered directly into the periaqueductal gray. However, injecting levorphanol into the amygdala of rats during passive avoidance conditioning, time- and dose-dependent decrease in retention was observed.[282] This effect could be reversed by naloxone. Supposing that the amygdaloid complex as a part of the limbic system plays a specific role in the learning process, we may conclude on a specific memory disrupting effect of opiates and on an unspecific facilitation of conditioning in certain experimental stiuations due to their rewarding (euphoric) action.

Being strong inducers of euphoria, opiates might have disinhibitory action too, i.e., they are expected to antagonize the conditioned inhibition. Accordingly, in operant conditioning paradigms pairing previously indifferent stimuli with painful electric shocks, an inhibition of learned behavior could be observed upon applying the conditional signal alone. This type of "conditioned fear"can be inhibited not only by the so-called "anxiolytic drugs",[283] i.e., by minor tranquilizers, but also by opiates.[284] In special situations animals may be compelled to compete for food or other reward. If the same animals are associated in such situations they form stable dominant-submissive hierarchy. Morphine was reported to disrupt the previously established hierarchy mainly by activating the submissive partner.[285] This is also a manifestation of their disinhibitory action.

There are numerous reports concerning the effects of opiates on operant behavior. Both facilitation and inhibition have been reported depending on the experimental variables. Opiates generally facilitate responding in low dosage and in schedules generating low base-line response rate, while opposite action can be observed if the dose or the base-line responding is already high.[286,287] These effects exerted on previously established conditional responding are probably connected with their action on motility. More or less similar data have been reported in relation to the active avoidance reactions. Previously elaborated avoidance reactions were found either suppressed[288,289] or facilitated,[290] while the acquisition of new conditioned reactions was unanimously reported to be inhibited[291,292] upon opiate treatment in such commonly used experimental situations as shuttle-box or pole-jumping tests. Just as in the case of classical operant responding, morphine was found to improve the performance at low dose levels and to diminish it at high dose levels.[289] Upon repeated application this biphasic dose-response curve was shifted to the right, i.e., tolerance developed to both the stimulant and depressant effects of morphine.[293] A further interesting variable of opiate effect is the individual learning capacity. Morphine facilitated acquisition of avoidance responding in subjects[294] and in strains[295] where the control performance was relatively low, but inhibited it at the "good responders". The difficulty of the task also influences the results. As reported by Ageel et al.,[296] for a given difficult task morphine enhanced acquisition at low doses, whereas in high dosage an inhibitory action was found. When increasing the task difficulty relatively larger doses were needed for both effects.[296]

The facilitation by morphine of avoidance behavior can be antagonized not only by opiate antagonists (as demonstrated in the above quoted studies) but also by α-methyl-p-tyrosine just as in the case amphetamine-facilitation.[290] Consequently, the improvement in performance might be due to opiate receptor mediated catecholamine release. An apparently strange hypothesis in view of the release inhibitory actions of opiates in vitro (Chapter 1, Volume I and Chapter 1, Volume II). However, as discussed previously, opiates cause a general locomotor facilitation upon dissipation of the initial depressant effects and probably hyperfunction of catecholaminergic neurones is involved in this late hyperactivity (Chapter 2, Volume I). Considering that elevating the activation level results in better performance but beyond an optimal point further activation detoriates learning (see the corresponding experimental psychological textbooks), these data might be related to facilitation of the catecholaminergic functions.

IX. MODULATION OF THE TURNOVER OF NEUROTRANSMITTERS AND SOME OTHER SUBSTANCES

Morphine-like drugs elicit very diverse biochemical effects in the brain and in the peripheral tissues as well. These metabolic alterations are to a great extent secondary consequences of changes in the normal homeostasis of the organism, i.e., they may be related to the stressor effect of opiates or to the compensatory processes, which follow the direct receptor mediated actions of opioids. Of their primary metabolic effects those taking place in the central nervous system deserve special attention since they may explain many behavioral actions. Of course all the neurotransmitters are of special interest and thus the data concerning them are reviewed separately. As all CNS drugs modify the turnover of neurotransmitters, especially those in the brain, the question arises whether any reliable information can be obtained on the mode of opiate actions upon recording the "unspecific" metabolic effects. The probable answer is yes, if they meet the main criteria of opiate specific actions, such as reversal by naloxone-like antagonists, development of tolerance upon repeated application, and similarity in effects of opiate agonists of different structures. Nevertheless, sometimes it is difficult to find out whether the biochemical and behavioral changes induced by narcotics represent two kinds of events independent of each other or the behavioral effects are only visible manifestations of the metabolic effects, both derivable from the stimulation of opiate receptors.

A. Effects on Acetylcholine (Ach) Turnover

The inhibition by morphine of stimulation-induced Ach release in guinea pig ileum and in other isolated organ preparations is a well-known phenomenon (Chapter 1, Volume I). Thus, it was reasonable to investigate if opiates exert a similar effect on the central cholinergic mechanisms. The findings reported from very different laboratories are on the whole unanimous.

As Table 13 shows, an increase in brain Ach content was observed in rats,[297,301] in mice,[238,303] and in cats[314] as well. Similar effects can be induced also by levorphanol but not by its optically inactive stereoisomer (dextrorphan).[303,304] Moreover, methadone also showed this effect.[314] Since the latter is structurally unrelated to morphine, this effect is connected with the opiate agonist activity. (The only exception might be meperidine, which did not exert statistically significant inhibition in one study, but this anomalous finding was attributed to the low dosage by the authors themselves.[314] The partial agonists have similar activity[238] while naloxone has not, but the latter effectively antagonized the increase in Ach content brought about by opiate agonists in the above-mentioned studies. Upon repeated administration tolerance develops to this effect of opiates.[303] Consequently, this change in brain Ach metabolism meets the main criteria of opiate specific actions.

Another question is how to interpret this elevation of brain Ach content. As the brain Ach is located in two separate pools within the neurones, i.e., a part of it is stored in synaptic vesicles while the other is not, the possibility of intracellular redistribution should also be counted with. However, both the "bound" and "free" fractions were found augmented[301,303,314] and their proportion was not changed. Thus, the increment must be explained in another way. An elevation of brain Ach content might be the consequence either of the accelerated synthesis or the decreased metabolism, or it might reflect an inhibition of release. In the first case similar effect can be expected in vitro. But no increment in Ach synthesis was observed in rat brain slices[306] and in no other studies did morphine and its derivatives modify significantly the Ach biosynthesis.[315,316] Hitherto nobody found any alteration of acetylcholine esterase activity upon morphine treatment, only in vitro in very high concentrations.[310]

Table 13
EFFECTS OF OPIATES ON ACETYLCHOLINE (ACH) METABOLISM IN VIVO

Treatment	Species	Effect	Ref.
Morphine	Rat	Increase in Ach content of the brain	297
Morphine	Cat	Decrease in Ach release from the cerebral cortex	298
Morphine	Mouse	Increase in brain Ach content, no change in cholinesterase activity	299
Morphine (locally over the cerebral cortex)	Cat	Decrease in Ach output into the lateral ventricles	300
Morphine	Rat	Increase both in free and bound Ach contents of the brain	301
Morphine	Rabbit	Decrease in epidural Ach output	302
Morphine	Mouse	Increase in bound Ach content of the brain	238
Pentazocine	Mouse	Increase in bound Ach content of the brain	238
Cyclazocine	Mouse	Increase in bound Ach content of the brain	238
Nalorphine	Mouse	Increase in bound Ach content of the brain	238
Morphine	Mouse	Increase both in free and bound Ach contents of the brain	303
Levorphanol	Mouse	Increase both in free and bound Ach contents of the brain	303
Dextrophan	Mouse	No effect	303
Morphine	Cat	Decrease in cortical Ach output	304
Levorphanol	Cat	Decrease in cortical Ach output	304
Dextrophan	Cat	No effect	304
Meperidine	Cat	Decrease in cortical Ach output	305
Methadone	Cat	Decrease in cortical Ach output	305
Pentazocine	Cat	Decrease in cortical Ach output	305
Nalorphine	Cat	Decrease in cortical Ach output	305
Morphine	Rat	Increase in cortical Ach content, no change in Ach syntesis	306
Morphine	Rat	Decreased choline acetyl transferase activity in caudate nucleus, no change in the cerebral cortex and thalamus	307
Morphine	Rat	Decrease in brain Ach turnover	308, 309
Morphine	Rat	Decrease in cortical Ach output	310
Morphine	Rat	Decrease in Ach turnover in the cerebral cortex but no change the striatum	311
Morphine (locally over the cerebral cortex)	Cat	No effect	312
Morphine	Cat	Decrease in cortical Ach output at small dose level and increase at high doses	313
Morphine	Cat	Increase in Ach content of the cortex, hypothalamus	314
Methadone	Cat	Increase in Ach content of the cortex, hypothalamus and midbrain	314
Meperidine	Cat	No effect	314
Morphine	Rat	Decrease and increase in Ach turnover in the periods of hypo- and hyperactivity, respectively	115

Consequently, the observed increase in brain Ach content may only reflect reduced liberation from the storage sites. Indeed many papers have been published on opiate-induced reduction of Ach output in vivo. This release inhibiting effect has been observed in mice,[303] in rats,[310] in rabbits,[302] and in cats[304,305] as well. A simple explanation of this action could be a direct inhibition of Ach synthesis. However, morphine caused only a 10% inhibition in a whole brain extract of the rat in the heroic concentration of 10^{-3} mole.[316] Moreover, both levallorphan and levorphanol had opposite effects in this experiment.[316] Accordingly, the decrease in Ach turnover rate cannot be attributed to a direct inhibition of its synthesis in the presence of opiates. It is worth mentioning that naloxone not only reverses the release inhibitory action of opiates, but it even increases the Ach output above the control level if it is given immediately after morphine.[304] It is probably due to the easy availability of Ach accumulated in its presynaptic stores upon opiate treatment. Thus, exaggerated Ach output upon naloxone treatment of morphine-dependent animals may be a component of the acute opiate withdrawal syndrome. Considering the release inhibitory effect of opiates in isolated organ preparations (Chapter 1, Volume I and Chapter 1, Volume II) and their incapability to inhibit its synthesis in brain slices,[306] the conclusion is simple. Inhibition of Ach liberation from the presynaptic storage sites results in increased steady-state level, since the vesicles are less frequently emptied and the compartment where Ach is accumulated is not accessible to cholinesterase. Consequently, the netto result is an increased accumulation due to the decreased rate of liberation and utilization. This is, however, tantamount to reduced Ach turnover. And really, inhibiting the in vivo Ach synthesis by i.c.v. given haemicholinium-3, its depletion was found slowed down upon pretreatment with morphine, methadone, propoxyphene,[308,309] an effect reversable by narcotic antagonists.[309] Thus modification of depletion in the presence of synthesis inhibitors is a sensitive indicator of turnover rate in the case of Ach too. Comparing the different morphine congeners, their relative activities were found similar to their relative analgesic potencies (heroin > levorphanol > methadone > phenazocine > morphine > codeine > meperidine) on i.p. treatment.[317] Although it is not clear whether the decreased Ach turnover is the cause of opiate analgesia or they are only concomitant events both due to activation of opiate receptors, in the case of morphine derivatives this metabolic effect is a reliable indicator of their pharmacological potency.

Interpreting the significance of these findings, the curious data of Howes et al.[238] are worth considering. These authors[238] found correlation between the whole brain Ach content and analgesic effect (in tail flick test) not only with morphine but also with oxotremorine. The latter's analgesic and brain Ach content increasing effects could also be antagonized by naloxone, nalorphine, and reserpine proving the similarities in their actions.[238] Regarding that oxotremorine is a postsynaptic muscarinergic receptor agonist, its naloxone-sensitive analgesic action is very difficult to explain.

Moreover, to state categorically that opiates inhibit Ach release from all nerve-terminals would be an unjustified oversimplification. Again the activation level, the dose applied, the species used, the brain area examined, and the time elapsed since opiate administration are factors which must be taken into consideration. As shown in Table 13 release-inhibitory effects have been uniformly reported in all rodent species examined. However, in cats (the only feline species frequently used in pharmacological laboratories) the data are more complex. Some authors found a decrease in Ach output[304,305] while others reported on biphasic action.[313,314] Since the cortical Ach output is known to correlate with the functional state of the brain and the general level of activation, the above-mentioned controversy in data might be due to the fact that not only conscious animals have been used in these studies. Ach output has been measured in several cases in anesthetized animals (see below), but it is well known that anesthetic agents reduce the Ach release and the effects of different drugs on Ach liberation is greatly influenced by the depth of anesthesia.[314] As for the influence of morphine on Ach output in cats, decrement was found in anesthetized animals.[304,305] Thus more reliable data might be obtained in pretrigeminally transsected[313] or "encéphale isolé"[314] preparations, where immobilization of the animals (needed in order to collect cerebrospinal fluid for Ach determination) does not necessitate anesthesia. Both groups found an inhibition of Ach release in small doses and facilitation at higher dose levels.[313,314] Moreover, the biochemical effect correlated well with the activation level upon morphine treatment: reduction was found in cats which were depressed or sleeping and an increase if they were alert or excited.[314] This correlation with the activation level was also observed for methadone.[314] Thus, again the questions arise whether these effects are specific, i.e., mediated by opiate receptors, and if yes, whether the same receptors mediate the opposite actions or not. Reversal by opiate antagonists of the antirelease effect of morphine was demonstrated in all the above-mentioned laboratories studying the problem. Consequently, both effects seem to be elicited by opiate receptors which, however, does not necessarily mean that the opposite actions are brought about by the same receptor population. Of course the increase in Ach output might be a compensatory process ensuing after the initial specifically induced decrease in liberation. Surveying all the publications concerned (Table 13) it turns out that almost all authors found only a monophasic change in the Ach metabolism. The observed effect reached its maximum generally 15 to 30 min after the i.v. administration of opiates and dissipated in 60 to 120 min. Postdepressant overshoot has not been reported. But according to Domino et al.[115] the morphine elicited subsequent phases of locomotor inhibition and facilitation are accompanied by concomitant decrease and increase in the Ach turnover, respectively. Therefore, the role of Ach in generation of the late excitatory behavioral symptoms remains to be elucidated.

A further important issue is the relative contribution of individual brain areas to the above-mentioned effects, which refer generally to the whole brain or mainly to the cerebral cortex. If the Ach metabolism of the brain is examined, generally the superfusion technique is applied, i.e., the artefical liquor is mostly collected over the surface of cerebral cortex for methodological reasons. When collecting the cerebrospinal fluid at the level of cerebral cortex, the observed effects will also be determined primarily by this structure. In order to study the Ach turnover in vivo in discrete brain areas, special techniques ought to be developed. This difficult task has been solved by Zsilla et al.[318,319] who infused labeled choline i.v. By determining the Ach content in different brain areas, the local turnover rate of Ach could be calculated. In these experiments the morphine congeners were found to modify the Ach turnover in different brain areas differentially.[264,319-321] The turnover decreased in the cerebral cortex, hippocampus, and n. accumbens if injecting morphine or meperidine systemically, but no effect was found in the striatum, n. septi, amygdala, substantia nigra, and n. raphe dorsalis.[264,319-321] That is, no effect was found in several areas rich in opiate receptors (Chapter 1, Volume III).

As for the striatum, Goldstein[321] found decrement in its Ach metabolism while others did not.[264,311] The lack of effect on striatal Ach turnover upon opiate treatment is a seemingly startling finding, since this structure is probably the richest not only in opiate receptors but also in cholinergic nerve terminals. Moreover, the rate of the Ach turnover of striatal neurones is several times higher than that of the cortical ones.[311] Cheney et al.[311] supposed that this apparent "noninvolvement" of striatal neurones was actually the netto effect of two opposing processes, i.e., morphine congeners not only directly inhibit the Ach release, but by a simultaneous suppression of dopamine liberation they abolish the tonic inhibition of cholinergic activity exerted by the former transmitter.[311] Experiments of Vizi et al.[322] give an in vitro confirmation of this hypothesis. These authors[322] found a clear-cut potentiation of ouabain-induced Ach release in striatal slices of rats by addition of morphine to the bathing solution. Naloxone, given alone, had no effect, but reversed that of the morphine.[322] Finally in this context it is worth recalling that striatal lesions did not modify significantly the morphine-induced analgesia and catatonia and microinjection of opioid substances into this structure also proved to be ineffective (Chapter 2, Volume I).

However, positive correlation has been found between analgesia and Ach turnover if the cerebral cortex and hippocampus were studied.[318] In these structures a decrease in Ach turnover was observed when applying the analgesic doses of morphine, azidomorphine, or meperidine;[318] a finding which may be surprising considering that the opiate receptor content of the hippocampus is rather low (Chapter 1, Volume III). Interestingly the easiest way to modify the hippocampal Ach metabolism by opiates was to inject them into the septum.[264,320] The latter area contains opiate receptors in high concentration (Chapter I, Volume II) and it has direct connections with the hippocampus.[264,320] Consequently, opiates might deccelerate the hippocampal Ach release via activating the septal opiate receptors.

A further yet unanswered question is whether morphine acts directly or indirectly on the cortical neurones. In several studies Ach release was suppressed upon local perfusion over the cerebral cortex[302] or through the lateral ventricles.[300] In another study, however, morphine failed to inhibit the liberation upon local application.[312] When relating the effect of opiates on cortical Ach turnover to their general behavioral actions, the former finding seems to be more "logical". The level of cortical activity is primarily governed by the ascending pathways connecting the midbrain reticular system with aspecific thalamic nuclei and cerebral cortex. As discussed above, the cortical Ach liberation is closely correlated with the activation level even upon morphine treatment. Regarding that the cortical activity is dependent on the aspecific ascending activating system, it is logical to assume that the decrease (or increase) in cortical Ach output is correlated with oscillation of the general activation level brought about by subcortical mechanisms, since the midbrain reticular formation contains opiate receptors in high density contrary to the cortex (Chapter 1, Volume III). Beleslin and Polak[300] suggested that the inhibitory effect of morphine, similarly to that of the general anesthetics on the cortical Ach release, was brought about by depression of cholinergic internuncials. As interneurones are particularly susceptible to the action of anesthetics,[300] the ultimate site of depressant action of morphine may also be on these interneurones.

All the above quoted authors agree that opiates are not capable of suppressing completely the Ach release either in vivo just as they inhibit only partially its release in the isolated organ preparations specifically sensitive to opiates (Chapter 1, Volume I). And similarly, under in vitro conditions the inhibition by morphine of the Ach release from the rat brain slices was found incomplete even in the heroic dose of 10^{-3} mole.[323] Thus, both in vitro and in vivo a part of the cholinergic activity is resistant to the inhibition mediated by the opiate receptors.

An even more complex picture emerges when trying to interpret the data of Yaksh and Yamamura,[324] who collected Ach through a cannula from the head of caudate nucleus of cats. Morphine depressed both the resting release and that induced by local electrical or chemical stimulation, but only the latter effects could be antagonized by naloxone.[324] Accordingly, the inhibitory effect of opiates might be partially independent of the opiate receptors. This unusual finding was obtained exclusively in the striatum,[324] but this structure behaves peculiarly in other aspects too. The activity of choline acetyltransferase was found diminished in rat caudate nucleus in vivo but not in vitro.[307] The lack of any effect in vitro excludes the possibility of a direct effect; it might rather be a consequence of a negative feedback mechanism activated by the increased Ach level due to opiate treatment.[307] Moreover, an increased rate of choline uptake into the striatal synaptosomes and a concomitant increase in Ach release was observed if the in vitro preparation was made from rats previously treated with morphine.[325,326] This phenomenon indicates increased Ach turnover in vitro. Consequently, the above discussed data on the decrease in Ach release from the cerebral cortex in vivo may not be particularly relevant in relation of the striatal mechanisms.

Table 14
EFFECTS OF CATIONS AND THEIR DEPLETION ON ANALGESIA INDUCED
BY MORPHINE-LIKE DRUGS

Treatment	Ion[a]	Species	Assay	Effect	Ref.
Morphine	Ca^{++}, Mg^{++}, Mn^{++}	Rat	Tail flick	Antagonism	26
Morphine	Sr^{++}, Ba^{++}, Ni^{++}, Cd^{++}	Rat	Tail flick	No effect	26
Morphine	Ca^{++}	Mouse	Electrical stimulation of the tail	Antagonism	328
Meperidine	Ca^{++}	Mouse	Electrical stimulation of the tail	Antagonism	328
Morphine	EDTA	Mouse	Electrical stimulation of the tail	Potentiation	328
Meperidine	EDTA	Mouse	Electrical stimulation of the tail	Potentiation	328
Morphine	Na^+ citrate	Mouse	Electrical stimulation of the tail	Potentiation	328
Meperidine	Na^+ citrate	Mouse	Electrical stimulation of the tail	Potentiation	328
Morphine	Mg^{++}, Ba^{++}, Zn^{++}	Mouse	Electrical stimulation of the tail	No effect	328
Morphine	K^+, Na^+, Al^{+++}	Mouse	Electrical stimulation of the tail	No effect	328
Morphine	La^{+++}	Mouse	Tail flick	Potentiation	330
Morphine	Ca^{++}	Mouse (morphine dependent)	Tail flick	Hyperalgesia	331
Morphine	EGTA	Mouse	Tail flick	Potentiation	331

Note: EDTA: ethylenediaminetetraacetic acid; EGTA: ethyleneglycol-*bis*-(β-aminoethyl ether)-*N,N'*-tetraacetic acid

[a] Given intracerebrally.

A further interesting finding obtained in anesthetized rats is that the morphine-induced inhibition of Ach release from neocortex could be reversed by s.c. administered $CaCl_2$, but the latter alone did not modify the liberation.[327] Consequently, the suppressant action of morphine is mediated through an interaction with the calcium ions. In other studies,[328,329] the antagonism by calcium and some other divalent cations of opiate-induced analgesia has been reported (Table 14). Therefore, calcium depletion was supposed to play an important role in both the metabolic and analgesic effects of opiates,[327-329] since according to Jhamandas et al.[332] the inhibition by morphine of cortical Ach release could be antagonized by i.c.v. administered calcium or i.v. injected teophylline. The latter's effect might be attributed to its calcium mobilizing and phosphodiesterase inhibitory activities (see also Table 17).

B. Effect on Serotonin (5-HT) Turnover

Taking into consideration the results of drug interaction studies (Tables 13, 15, 16, 17 and 20 in Chapter 2, Volume I), the inhibition of opiate analgesia by 5-HT depletion (Chapter 2, Volume I) or by destruction of brain regions containing 5-HT neurones (Chapter 2, Volume I) and the latters' correlation with the brain structures from where antinociception can be elicited by microinjection of opiates (Chapter 2, Volume I), the turnover of this neurotransmitter is expected to be specifically modified by morphine-like drugs. Fortunately, as for the role of 5-HT in opiate analgesia, the data available are almost unanimous or at least less controversial than in relation of Ach turnover.

The first data on the role of 5-HT liberation in the action of opiates were demonstrated in a peripheral (*in situ*) preparation. Burks and Long[333] detected 5-HT release into perfused vasculature of isolated dog intestinal segments after intra-arterial administration of morphine, meperidine, levorphanol, nalorphine, codeine, and also apomorphine. If the dogs were pretreated with reserpine, no 5-HT release was detected in the venous effluent of the locally perfused intestinal segment.[333] The antagonism by reserpine of opiate-induced 5-HT prove that the latters do not act postsynaptically but on the presynaptic side.

With the exception of one study[334] no report has been published on elevation of brain 5-HT level upon treatment with opiates. Nevertheless, due to its very rapid degradation the steady-state level cannot be used to measure its turnover. Numerous reports have been published, however, on the increase in 5-HT turnover upon morphine treatment (Table 15). Interestingly, in the very first study[326] 5-HT turnover was found unaltered upon morphine treatment in contrast to the latter reports. It is conceivable that this negative finding was due to certain technical difficulties.[326] The best indicator of 5-HT metabolism is probably the accumulation of 5-hydroxyindolacetic acid (5-HIAA) which is its principal metabolite. To measure reliably the 5-HIAA level, the animals must be pretreated with agents (generally probenicid is used) which inhibit the former's clearance from the brain. Way et al.,[326] however, calculated the 5-HT turnover by measuring its accumulation in animals pretreated with a monoamine oxydase inhibitor (pargyline). Probably the latter did not block completely the degradation of 5-HT rendering the estimation less sensitive.[326] Blocking, however, the amino acid transport system by probenicid, the 5-HIAA accumulation was found linear[336] proving that it is a valid indicator of 5-HT synthesis. First Yarborough and his co-workers demonstrated the increase in 5-HT metabolism in the rat[335,337] and in the mouse[336] brain as well. After a single injection of 16 to 32 mg/kg morphine the 5-HIAA level was significantly elevated with the peak effect occurring at 2 hr.[337] At higher dose levels both the duration and the magnitude of the change were increased.

As for the contingent correlation between increased serotonin turnover and morphine-induced analgesia, it seems to be very close. In experiments made by Haubrich and Blake[338] in rats, two diagrams one showing the brain 5-HIAA level and the other displaying the changes in pain threshold, ran parallel, i.e., by gradual normalization of brain 5-HT turnover also the antinociceptive effect dissipated. Also this metabolic effect of morphine can be antagonized by naloxone.[341,346]

In one study[339] other opiates such as pentazocine, methadone, and meperidine were not found to increase the brain 5-HT turnover; a curious finding, which requires further confirmation. Of course the synthetic morphine derivatives are known to also differ qualitatively from morphine, but activation of the serotoninergic functions is such a basic property of opiates that this finding[339] is puzzling anyway.

Increased turnover of a brain transmitter may reflect either postsynaptic receptor blockade and/or its increased presynaptic output. As there is no reason to suppose a serotoninergic blockade by opiates, the former explanation has long been accepted but only recently demonstrated.[271,348]

Aiello-Malmberg[348] and her colleagues demonstrated tremendously increased 5-HT release from the cerebral cortex and caudate nucleus in brain stem transected cats and from the cerebral cortex of anesthetized rats upon morphine treatment. In the same study, the direct electrical stimulation of n. linearis intermedius of raphe in cats also augmented the 5-HT release from the cortex and the caudate nucleus as well.[348] Considering that the majority of 5-HT releasing nerve terminals in the forebrain originate from cell bodies located in the raphe nuclei, the conclusion is evident: morphine activates the ascending serotoninergic pathways. But the same nuclei represent also the starting point of the descending serotoninergic pathways and their involvement in op-

Table 15
EFFECTS OF OPIATES ON SEROTONIN (5-HT) METABOLISM IN VIVO

Treatment	Species	Effect	Ref.
Morphine	Dog	Increased release of 5-HT from intestinal vasculature	333
Levorphanol	Dog	Increased release of 5-HT from intestinal vasculature	333
Methadone	Dog	Increased release of 5-HT from intestinal vasculature	333
Meperidine	Dog	Increased release of 5-HT from intestinal vasculature	333
Nalorphine	Dog	Increased release of 5-HT from intestinal vasculature	333
Morphine	Mouse	No change in 5-HT turnover of the whole brain	326
Morphine	Rat	Increase in 5-HT turnover of the whole brain	335
Morphine	Mouse	Increase in 5-HT turnover of the whole brain	336
Morphine	Rat	Increase in 5-HT turnover in the brain stem	185
Morphine	Rat	Increase in 5-HT turnover but no change in the tryptophan hydroxylase activity in the whole brain	337
Morphine	Rat	Increase in 5-HT turnover but no change in its steady-state level in the brain	338
Morphine	Rat	Increase in 5-HT turnover of the brain	339
Methadone	Rat	No effect	339
Pentazocine	Rat	No effect	339
Meperidine	Rat	No effect	339
Morphine or heroin	Rat	Increase in brain 5-HT level and turnover, increase in tryptophan hydroxylase activity	340
Morphine	Rat, mouse	Increase in brain 5-HT turnover but no change in its steady-state level	341
Morphine	Rat	Increased conversion of 5-hydroxytryptophan into 5-HT	342
Morphine	Rat	Accelerated conversion of tryptophan into 5-hydroxytryptophan and serotonin	343
Morphine	Rat	No effect on synaptosomal 5-HT uptake	344
Morphine	Rat	Increased penetration of tryptophan from the plasma into the brain	345
Morphine	Rat	Increase in spinal 5-HT turnover	346
Morphine	Rat	Considerable increase of 5-HT turnover in lumbar cord, medulla, pons and to smaller extent in the midbrain and forebrain	102
Morphine	Rat	Increase in spinal 5-HT turnover	347
Morphine	Mouse	Increase in brain 5-HT and 5-HIAA levels	334
Morphine into the periaqueductal gray	Rat	Increase in 5-HT level in the spinal cord	271
Morphine	Rat	Increase in 5-HT release from the cerebral cortex	348
Morphine	Cat	Increase in 5-HT release from the cerebral cortex and caudate nucleus	348

Note: 5-HIAA: 5-hydroxyindol acetic acid; 5-HT turnover assessed by the 5-HIAA level.

iate analgesia has been demonstrated in various ways (Chapter 2, Volume I and Chapter 8, Volume III). Actually, the specific role of ascending rather than that of the descending serotoninergic pathways requires still further clarification. A further interesting finding is the facilitation by physostigmine of 5-HT release.[347] In the light of this finding the analgesic and opiate analgesia potentiating effects of cholinergic stimulants are easy to understand. (Chapter 2, Volume I) Moreover, the incapability of naloxone-like agents to reverse the analgesia induced by cholinergic drugs[235] may also be explained if supposing that antinociception is mediated directly by serotoninergic mechanisms which may be activated both by opiate and 5-HT receptors.

Recently, several reports dealt with the effect of morphine on 5-HT turnover in discrete parts of the rat brain. Obviously those regions may be involved in the opiate-induced activation of serotoninergic mechanisms where the 5-HIAA (and serotonin) levels are significantly enhanced upon opiate treatment. This site seems to be first of all the spinal cord.[102,271,346,347] Moderate increment was found in the pons and the medulla while the smallest was found in the midbrain and forebrain.[102] The facilitation of 5-HT turnover in the spinal cord upon opiate treatment[102,271,346,347] shows the importance of descending serotoninergic pathways in opiate analgesia. According to Shiomi et al.[346] the 5-HIAA content increases in both halves of the spinal cord upon opiate treatment but only the concentration measured in the dorsal half of the spinal cord, i.e., where the descending serotoninergic fibers are located, correlates with morphine analgesia in its time-course. Intercollicular transection did not modify the morphine-induced elevation of spinal 5-HIAA content, but spinalization at the C_1 level did.[346] Transecting the spinal cord at midthoracic level, increased 5-HIAA concentration could be detected above the transection level but not below it.[346] Yaksh and Tyce[271] injecting morphine into the PAG observed 5-HT liberation in the spinal cord concomitantly with analgesia. Let us mention here that direct intrathecal application of 5-HT also resulted in analgesia[349,350] and the analgesic effect of morphine injected into the PAG could be inhibited by intrathecally administered methysergide.[272] Deakin and Dostrovsky[351] injecting 5,6-dihydroxytryptamine into the dorsal raphe nucleus, i.e., irreversibly depleting the forebrain 5-HT content, found the analgesic effect of morphine essentially unaltered. However, intrathecal application of this selective 5-HT depleting agent significantly attenuated the antinociceptive effect of morphine as determined in the tail flick test.[351] As it is discussed in Chapter 2, Volume II iontophoretically applied 5-HT antagonized the dorsal horn unit activity evoked by nociceptive stimuli. Thus, the analysis of biochemical changes induced by opiates lends further support to the hypothesis postulating that the descending serotoninergic pathways play decisive roles in opiate analgesia.

But almost nothing is known as to the mechanism of facilitation of 5-HT turnover upon opiate treatment. One theoretical possibility is the inhibition of its uptake into synaptosomes. As a matter of fact, morphine was reported to inhibit the 5-HT reuptake into the hypothalamic synaptosomes in vitro.[344]

Neither acute nor chronic morphine treatment had any similar action, and even the inhibitory effect of in vitro added morphine was unaffected by pretreatment with it in vivo.[344]

A further possibility is an increment in the rate of conversion of tryptophan to 5-hydroxytryptophan and to 5-HT, i.e., an augmentation of tryptophan hydroxylase activity. The increased rate of these first steps of 5-HT metabolism has been repeatedly demonstrated.[337,340,342] This increment may be either the primary effect of opiates or simply a secondary consequence of increased utilization due to its facilitated release. Hypothetically a drop in the intracellular level of 5-HT should lead to its increased synthesis via the loss of end-product mediated inhibition. So far no data have been published which would confirm such an explanation. Moreover, Yarbrough et al.[337] have observed no alteration in the in vitro activity of tryptophan hydroxylase after a single injection of morphine. Consequently, its facilitatory effect on 5-HT turnover does not seem to be explained by a direct effect on the activity of the rate-limiting enzyme. Accordingly, a further contingency must be counted with: probably the brain level of tryptophan is increased upon opiate treatment. Indeed in one study the brain tryptophan level was found elevated after morphine treatment and this effect could be antagonized by naloxone.[345] Others, however, did not found the brain tryptophan level increased upon morphine administration.[337] A further possible way of action is a facilitated extraction of this amino acid from the plasma since its passage through the blood-brain barrier is an active therefore saturable process. However, its uptake was found even decreased in some studies[342,343] or just unaltered by others.[340] According to Larson and Takemori[342] the increased rate of 5-HT turnover upon acute morphine treatment is due to the increased rate of 5-hydroxytryptophan uptake and its facilitated conversion to 5-HT. Since the tryptophan uptake from the blood was actually decreased, the increased availability of tryptophan was attributed to the general inhibition of protein synthesis elicited by morphine.[342]

C. Effect on Catecholamine Turnover

Though dopamine (D), noradrenaline (NA), and adrenaline (A) have distinct, in certain aspects, even opposite functions in the brain and in the autonomic nervous system, the common discussion of the effects of morphine on their metabolism might be justified by the fact that the first steps in their synthesis are the same. Therefore, their metabolism might be influenced in similar way upon opiate treatment.

First Vogt[352] reported in 1954 that acute administration of morphine caused a decrease in the NA concentration in the cat brain. This finding was later confirmed and extended, namely a similar decrease was observed in rabbits, dogs, and rats.[353] As a concomitant decrease of A concentration was detected in the adrenal medulla, these findings were interpreted as manifestations of the general activation of sympathetic nervous system regarding the noradrenergic neurones of the brain a part of it. This might really be the case in those studies where enormously high doses of morphine were applied and the catecholamine levels were determined 5 hr after the drug administration. In this study of Maynert and Klingman[353] the stressor effect of opiates and/or the late compensatory processes could have confounded the primary specific action.

Nevertheless, also the primary effect of morphine on catecholamine metabolism is connected with an increased rate of turnover. This effect was demonstrated mostly by measuring the decline of brain D and NA levels upon blocking the tyrosine hydroxylase activity by α-methyltyrosine (AMT). Due to the increased impulse flow along the catecholamine liberating fibers D and NA depletion was accelerated as described first by Gunne et al.[354] and confirmed subsequently by many other groups (Table 15).

In the process of catecholamine synthesis the first step is the rate limiting one, i.e., conversion of tyrosine to DOPA. Injecting labeled tyrosine into the ventricular space in the rat brain, increased accumulation was observed by Clouet and Ratner[355] if the animals were pretreated with morphine. This observation, repeatedly confirmed in the subsequent studies in rats and in mice as well (Table 15), indicates the facilitation of tyrosine hydroxylase activity in vivo. However, the in vitro activity of this enzyme in mouse brain homogenate was not significantly influenced by morphine even in very high concentrations.[358] Thus, the increment in the catecholamine turnover rate cannot be attributed to a direct action of opiates exerted on it just as in the case of 5-HT metabolism (see above).

Otherwise, in the subsequent studies increased brain homovanillic acid (HVA) and 3,4-dihydroxyphenylacetic acid (DOPAC) levels were detected in mouse whole brain and rat striatal homogenates.[149,357] These classical experiments made by Fukui and Takagi,[357] Kuschinsky and Hornykievicz,[149] also repeated by many others, show that mainly D is the end-product of the catecholamine turnover facilitated by opiates, since HVA and DOPAC are its principal metabolites in the brain tissue.

These metabolic effects meet the main criteria of opiate specific actions; they may be reversed by opiate antagonists[352,356,361,364,368] and similar activity is displayed by other opiate agonists such as methadone, levorphanol, phenazocine, and pentazocine but cyclazocine of predominantly antagonistic character has no such effect.[361,364]

The next intriguing question is that which of the brain catecholamines are primarily involved in these changes. Data summarized in Table 16 show that in opiate-treated animals NA accumulation may be observed mainly in the lower parts of the brain

Table 16
EFFECTS OF OPIATES ON CATECHOLAMINE METABOLISM IN VIVO

Treatment	Species	Effect	Ref.
Morphine	Rat	Decrease in brain NA and adrenal A contens	352
Morphine	Dog	Decrease in brain NA and adrenal A contens	353
Morphine	Rat	Decrease in brain NA and adrenal A contens	353
Morphine	Rabbit	Decrease in adrenal A but no change in brain NA contents	353
Morphine	Rat	Accelerated depletion of brain dopamine after AMT treatment	354
Morphine	Rat	Accelerated conversion of labeled tyrosine to dopamine and NA in the brain	355
Morphine	Mouse	Accelerated conversion of labeled tyrosine to dopamine and NA in the brain	356
Morphine	Mouse	Increase in the whole brain concentrations of DOPAC and HVA	357
Morphine	Rat	No effect on steady-state concentrations of NA and dopamine in the brain stem	185
Morphine	Mouse	Accelerated conversion of labeled tyrosine to dopamine and NA in the brain	358
Morphine	Rat	Increase in the striatal HVA level	149
Morphine	Rat	Accelerated conversion of labeled tyrosine to dopamine in the striatum	359
Morphine	Mouse	Increased tyrosine hydroxylase activity in vitro after in vivo treatment	360
Morphine	Rat	Accelerated dopamine depletion upon AMT treatment mainly in the striatum	361
Methadone	Rat	Accelerated dopamine depletion upon AMT treatment mainly in the striatum	361
Pentazocine	Rat	Accelerated dopamine depletion upon AMT treatment mainly in the striatum	361
Morphine	Rat	Accelerated depletion of dopamine in the forebrain and of NA in the pons and medulla upon AMT treatment	75
Fentanyl	Rat	Initial decrease followed by elevation of the striatal HVA content	362
Morphine	Rat	Accelerated depletion of dopamine in the frontal cortex, but not in the striatum upon AMT treatment	363
Morphine	Mouse	Increased accumulation of newly synthesized dopamine and NA in the whole brain	364
Methadone	Mouse	Increased accumulation of newly synthesized dopamine and NA in the whole brain	364
Levorphanol	Mouse	Increased accumulation of newly synthesized dopamine and NA in the whole brain	364
Morphine	Rat	Increase in DOPAC and HVA levels in the n. accumlens, striatum and olfactory tubercle	365
Morphine into the n. reticularis gigantocellularis	Rat	Increased normetanephrine level in the spinal cord but no change in the spinal NA concentration	366
Morphine	Mouse	Increased normetanephrine and 3-methoxytyramine levels in the spinal cord but not in the brain	367
Morphine	Rat	Accelerated dopamine and NA depletion upon AMT treatment in the brainstem but not in the cerebral cortex	368
Morphine	Rat	Acceleration of dopamine turnover in the striatum and its decceleration in the frontal cortex and median eminence	369

Table 16 (continued)
EFFECTS OF OPIATES ON CATECHOLAMINE METABOLISM IN VIVO

Treatment	Species	Effect	Ref.
Morphine	Rat	Increased efflux of dopamine from striatal neurones with lowered firing rate but no change if the firing is normal	370

Note: NA = noradrenaline; AMT = α-methyl-*p*-tyrosine; DOPAC = 3,4-dihydroxyphenylacetic acid; A = adrenaline; HVA = homovanillic acid.

stem,[75,368] while the D concentration was found elevated mostly in forebrain structures, first of all in the striatum.[75,149,359-362,365,369,370] Only one group[363] found no effect on D synthesis but an acceleration in the frontal cortex upon morphine treatment. The findings of Westerink and Korf[365] who observed more pronounced accumulation of D metabolites in n. accumbens than in the striatum and in olfactory tubercle deserve special attention. As for the spinal cord, its normetanephrine and 3-methoxytyramine contents were found elevated upon opiate treatment, while the concentrations of these metabolites remained the same in the brain.[367] Similarly to these findings of Kameyama et al.[367] obtained in mice, another Japanese group[366] detected an increase in the spinal normetanephrine level in rats. In the former study[367] morphine was injected systemically. Accordingly this peculiar change in the spinal catecholamine turnover might reflect changes not related to its analgesic effect. However, Kuraishi et al.[366] observed a similar elevation of spinal normetanephrine level upon microinjection of morphine into the n. reticularis gigantocellularis of the medulla oblongata, but no such change was produced by administering it into the PAG, n. raphe magnus, i.e., into other brain stem loci from where antinociception can be elicited by opiates[366] (Chapter 2, Volume I). Since normetanephrine is an extraneuronally produced metabolite of D and NA,[367] these data show the role of a noradrenergic medullo-spinal pathway in mediation of opiate analgesia. Moreover, these data are in good agreement with those of an other laboratory,[371] where intrathecal phentolamine was found to partially antagonize the analgesic effect of morphine given into the PAG or with the report of Yaksh,[272] who observed elevation of the nociceptive threshold upon spinal injection of adrenergic agonists.

The median eminence represents a further brain area where the effect of morphine on the catecholamine turnover is "anomalous". In this region a dose-dependent decceleration of D turnover was observed.[269] This action is also reversible by naloxone, i.e., mediated by opiate receptors. As the prolactin release is tonically inhibited by the tubero-infundibular D releasing neurones, suppression of D liberation by opiates may explain their prolactin secretion increasing action. In short, the data on the effects of opiates on brain catecholamine metabolism are essentially unanimous if certain regional differences are also taken into consideration, Unfortunately, for the time being no comprehensive interpretation of these actions can be given. Even more obscure is their relation to the behavioral and analgesic actions of opiates.

The morphine-induced decceleration of D turnover in the median eminence is probably the easiest to interpret if considering that in many in vitro preparations opiates inhibit the release of different neurotransmitters via presynaptic receptors (Chapter 1, Volume I). Reduced release generally causes metabolic inhibition by accumulation of otherwise regularly secreted end-products in the presynaptic stores. But how to explain the increased rate of D synthesis in other forebrain structures or that of NA in the brainstem?

First it has been suggested by Kuschinsky and Hornykievicz,[140,149] that morphine increases the intraneuronal D metabolism by displacing it from the storage sites to the sites of catabolism. However, no subsequent data have been reported which would confirm this hypothesis.

Either increased release or decreased reuptake into the nerve terminals may also lead to depletion and accelerated synthesis. In some studies inhibition of D[372] or NA[373] uptake has been reported but only in high doses and no correlation has been found between the uptake inhibition and analgesia. In subsequent experiments no such effect was detected at all.[374,375] Moreover, in the latter in vitro studies rather an inhibition of potassium-induced D release was observed in rat striatal slices and in the synaptosomal fraction of rat striatum.[375] In another study morphine inhibited the NA release from cortical slices but had no effect on striatal D release.[376] These findings are compatible with the general transmitter release inhibiting effects of opiates but they cannot explain the well-documented increase in D metabolism.

As we are confronted with a contradiction of the in vitro and in vivo obtained data, probably the formers may be disregarded. Namely it should be taken into consideration that the striatum is only a part of the complicated nigro-striatal circuitry with numerous inputs, built-in feedback mechanisms, and synapses working with different transmitters among others D, Ach, GABA, and enkephalin. Thus, it might be assumed that its artificial isolation, deprivation of its natural connections results in completely unreliable data in the customary in vitro experiments.

These findings may be interpreted more easily if thinking in terms of receptors. Due to their characteristic behavioral effects opiates have been supposed to inhibit the dopaminergic transmission. Strong sedation, catalepsy, increase in muscular tone, and the acceleration of striatal D turnover are characteristic actions of neuroleptics. Also a direct postsynaptic D receptor blocking action was suspected,[51] a working hypothesis which was seriously considered before the discovery of opiate receptors. However, the behavioral symptoms elicited by neuroleptics are qualitatively different (Chapter 2, Volume I), they do not cause analgesia, do not induce dependence, while the opiates do not inhibit unambiguously the apomorphine-induced stereotypy,[153-155] which is a very sensitive manifestation of postsynaptic D receptor blockade. In our own unpublished experiments morphine suppressed the apomorphine elicited gnawing only at high (nonspecific) dose levels (10 to 30.0 mg/kg), where the inhibition of stereotypy might be due to the general motor debilitating, muscle tone enhancing, and cataleptogenic actions. In the same experiments haloperidol significantly attenuated the apomorphine elicited stereotypy already in the minute amounts of 0.05 to 0.10 mg/kg, i.e., in doses which hardly induced any visible alteration in the spontaneous locomotion and muscle tone. The main evidence against a direct action of opiate agonists on the postsynaptic D receptors represent the differences in antagonizability: while the cataleptogenic action of neuroleptics can be inhibited by atropine, that of the opiates can be specifically antagonized only by naloxone-like agents (Chapter 2, Volume I).

A blockade of the presynaptic dopaminergic (auto)receptors, i.e., suppression of presynaptic feedback mechanisms also cannot explain the behavioral symptoms characteristic of decreased extrapyramidal activity. However, the acceleration of D turnover may really be attributed to the loss of presynapticaly (autoreceptor) mediated negative feedback regulation. Another line of evidence also points to a presynaptic site of action. Namely upon repeated morphine or methadone treatment the ability of amphetamine to induce stereotypy or that of apomorphine to deccelerate the D turnover was gradually enhanced, i.e., the postsynaptic D receptors became hypersensitive.[377] The supersensitivity of peripheral noradrenergic nerve terminals has also been demonstrated upon repeated opiate treatment.[378] The supersensitivity of postsynaptic receptors is frequently a secondary consequence of presynaptic inhibition of transmitter release. Accordingly, almost all the current hypotheses concerning the opiate tolerance and dependence development are based on presynaptic inhibition of neurotransmitter liberation resulting in disuse hypersensitivity of the postsynaptic membrane (Chapter 9, Volume III).

In a recent experiment of Andén and Andén-Grabowska,[379] rats with unilateral intrastriatal KCl injection responded by ipsi- or contralateral turning to apomorphine and haloperidol, respectively, and morphine antagonized both reactions. In the same report morphine did not modify the apomorphine elicited inhibition of D turnover.[379] The latter effect, i.e., decceleration of D turnover, is generally regarded as an autoreceptor mediated presynaptic process. The inability of morphine to inhibit this action of apomorphine indicates that it does not act via the presynaptic dopaminergic autoreceptors. On the other hand, both substantia nigra and striatum are extremely rich in opiate receptors and a considerable fraction of them are located on dopamine releasing nerve terminals (Chapter 1, Volume III). These morphological data point to a third possibility according to opiate receptor containing neurons located close to the D releasing neurones (in the substantia nigra) and in the vicinity of D sensitive postsynaptic membranes (in the striatum) transsynaptically inhibit the impulse transmission simultaneously at two different sites.

Moreover, it should be taken into consideration that destruction of the striatal structures does not significantly modify the morphine-induced catalepsy or opiate analgesia (Chapter 2, Volume I). Hence, the morphine-induced facilitation of striatal D turnover might have no relevance as for its behavioral and analgesic effects. However, Moleman and Bruinvells[370] recently reported that though morphine did not modify the striatal turnover in untreated or haloperidol treated rats, it reversed the deccelerating action of apomorphine. As similar effects could be observed in animals with decreased D turnover due to lesioning of the substantia nigra,[370] this effect was supposed to be mediated by the dopaminergic axons via opiate receptors. Finally in our laboratory[380] morphine (given s.c.) inhibited the turning behavior elicited by apomorphine or amphetamine in rats with unilateral electrolytic nigral lesion only at very high dose levels. Therefore, the data though heterogenous suggest that opiates do not act directly on the striatal dopaminergic receptors located either pre- or postsynaptically; their effects are indirect and complex mediated by independent neurones and receptors situated presumably to a large extent extrastriatally.

D. Effects on Dopamine Sensitive Cyclic Nucleotides

There is hardly a more controversial issue of the opiate research than their effects on cyclic nucleotides. Probably the main difficulty arises upon trying to reconcile the in vitro and in vivo data. As discussed elsewhere, (Chapter 1, Volume II) opiates inhibit the cAMP formation in vitro from ATP, a process catalyzed by membranal adenylate cyclase (AC). The resulting decrease in intracellular cAMP level may be observed both in brain slices and in opiate receptor containing neuroblastoma X hybrid cells (Chapter 1, Volume II). Morphine congeners are reported to vigorously inhibit the PGE_1-induced stimulation of AC activity, while an increase in the activity of the latter has been repeatedly suggested to initiate the tolerance and dependence development.[381] As the striatum is the brain area richest in D, mostly striatal slices have been used in the corresponding studies. Since AC is the second messenger in activation of catecholaminergic cells, the whole issue is connected with the effects of opiates on dopaminergic and adrenergic synaptic transmission.

As shown in Table 17, in vivo increased cAMP level has been found by several groups in the striatum,[385,390,409] while in other brain regions decrement or no significant

Table 17
OTHER IN VIVO METABOLIC EFFECTS OF OPIATES

Treatment	Species	Effect	Ref.
Morphine	Guinea pig	Accelerated incorporation of phosphorus into brain phospholipids	382
Morphine	Rat	Hyperglycemia	383
Morphine	Rat	Ca^{++} depletion in the striatum, hypothalamus and cerebral cortex	384
Morphine	Rat	Increase in cAMP level in caudate nucleus, decrease in hypothalamus, and substantia nigra and no change in thalamus	385
Morphine	Rat	Decrease in cGMP level in the caudate nucleus, hypothalamus, substantia nigra, and thalamus	385
Morphine	Rat	Increase in dopamine sensitive AC activity in the caudate nucleus	386
Levorphanol	Rat	Increase in dopamine sensitive AC activity in the caudate nucleus	386
Morphine	Mouse	Inhibition of uridine incorporation into the brain RNA but no similar change in the liver	387
Morphine	Rat	Ca^{++} depletion in 8 different brain areas	388
Morphine	Rat	Ca^{++} depletion in 8 different brain areas	388
Dextrorphan	Rat	No effect	388
Morphine	Rat	Decrease in cGMP content of the cerebellum	389
Morphine (intrastriatally)	Rat	Decrease in cGMP content of the cerebellum	389
Morphine	Rat	Increase in the striatal cAMP level, AC and PD activities	390
Morphine	Mouse	No effect on AC, PD, cAMP-protein kinase activities and on cAMP accumulation in cerebral slices	391
Morphine	Rat	No effect on histamine level in the hypothalamus	392
Morphine	Rat	Increase in GABA level and turnover in the anterior thalamus and in the spinal cord (dorsal horn)	197
Morphine	Rat	Depletion of the brain synaptosomal Ca^{++} but no change in Na^+, K^+, and Mg^{++} contents	393
Morphine	Rat	Decrease in cerebellar cGMP content	394
Dextromoramide	Rat	Decrease in cerebellar cGMP content	394
Morphine	Rat	Inhibition of lysine incorporation into brain proteins	395
Morphine	Rat	Increased rate of thiamine incorporation into the cortex, brainstem and cerebellum	396
Morphine	Mouse	Decrease in synaptosomal Ca^{++} content	397
Morphine	Mouse	Increase in plasma cAMP level	398
Morphine	Rat	Increase in the cortical but no change in the hypothalamic cAMP levels	400
Methadone	Neonatal mouse	Inhibition of uridine incorporation into brain RNA and that of leucine into brain proteins	401
Morphine	Rat	No change in GABA content of the striatum, midbrain, and cerebral cortex	402
Morphine	Rat	Decrease in synaptasomal Ca^{++} content and its increased binding to certain subfractions of the synaptic plasma membrane	403
Morphine	Rat	Increase in liver tyrosine aminotransferase activity	404
Morphine	Rat	Facilitation of prostaglandin synthesis in the placenta	406
Morphine	Mouse	No significant change in synaptosmal Ca^{++} efflux	407
Levorphanol	Mouse	No significant change in synaptosomal Ca^{++} efflux	407
Morphine	Mouse	Decrease in synaptosomal Ca^{++} uptake	408
Morphine	DBA/2J mouse	Decrease in cerebellar cGMP but no change in striatal cAMP levels	409

Table 17 (continued)
OTHER IN VIVO METABOLIC EFFECTS OF OPIATES

Treatment	Species	Effect	Ref.
Morphine	C57 mouse	Increase in cerebellar cGMP and in striatal cAMP levels	409
Morphine	Mouse	Increase in plasma cAMP level	399
Meperidine	Mouse	Increase in plasma cAMP level	399
Pentazocine	Mouse	Increase in plasma cAMP level	399
Morphine	Mouse	Inhibition of brain dopamine and NA sensitive Na^+-K^+-ATPase activity, no effect on brain Mg^{++} ATPase activity	410
Morphine	Rat	Increased synthesis of certain phosphoinositides in the midbrain	412
Morphine	Rat	Dose- and time-dependent biphasic effect on the brain histamine concentration	413
Morphine	Mouse	Increased rate of tryglyceride production in the liver	414
Morphine	Mouse	Increased activity of phosphatidate phosphohydrolase in the liver	414
Morphine	Mouse	Increase in serum glutamate oxalacetate transaminase activity	415
Morphine	Mouse	Increase in serum glutamate pyruvate transaminase activity	415
Morphine	Mouse	Increased rate of gluconeogenesis and ureogenesis in the liver	405

Note: AC = adenylate cyclase; NA = noradrenaline; PD = phosphodiesterase.

alteration have been measured. Puri et al.[390] reported also on the simultaneous increment in AC and phosphodiesterase activities, i.e., the increased level resulted in accelerated turnover of cyclic nucleotides. According to Iwatsubo and Clouet[386] both the basal and the D sensitive AC activities are enhanced upon morphine administration in the caudate nucleus, i.e., the per cent stimulation by D remained constant in spite of the increment in basal and D-stimulated activities. They assumed that the enhanced AC activity in the nerve terminals was due either to the activation of the enzyme *in loco* or to an accelerated transport from the sites of cyclase biosynthesis.[386] Addition of morphine to homogenized striatal tissue contrary to haloperidol did not modify the D-induced activation of AC.[386] Also these data show that morphine does not interact with the D receptors.

As shown in Table 17, similar effects may be obtained by levorphanol.[386] Also these in vivo effects may be reversed by naloxone,[385,409] consequently they are mediated by opiate receptors.

The elevated cAMP level and its accelerated turnover refer to an increased activity of the striatal dopaminergic system, a conclusion confirmed directly by Racagni et al.,[409] who compared the striatal metabolic effects of morphine in two different strains of mice showing different behavioral response to opiate treatment. One of the two strains (C57), known to display very intense running behavior upon morphine treatment also showed a definite increase in the striatal cAMP concentration, while in the other strain (DBA/2J), very sensitive to the analgesic action of morphine and displaying almost no locomotor stimulation upon treatment, the striatal cAMP level also failed to show significant elevation.[409] Consequently, the striatal AC activity is positively correlated with the locomotor activity but not with the analgesic effect. As for the cGMP level, it was found decreased in the cerebellum, concomitantly with the elevation of striatal cAMP concentration and also this effect is naloxone sensitive.[385,389,394,409] This is an apparently surprising finding in view of the virtual absence of opiate receptors in the cerebellum (Chapter 1, Volume III). Moreover, morphine failed to modify the guanylate cyclase or cGMP phosphodiesterase activities in cerebellar homogenates in vitro.[394] Thus Biggio et al.[394] suggested that facilitation of the striatal opiate receptors decreased the cerebellar cGMP level by reducing the activity of the mossy fiber excitatory input.

Returning to the cAMP, its concentration was found elevated also in the plasma,[398,399] an effect which is hardly related to the analgesic action but surely to the opiate receptors since it could be antagonized by naloxone. This increased level of plasma cAMP is presumably a manifestation of enhanced sympathetic tone and/or it is a part of the stress-response. The morphine-induced increase in plasma cAMP level was abolished by adrenalectomy, propranolol, or pentolinium but not by pretreatment with atropine, α-methyltyrosine, or phentolamine.[399] Hence, the phenomenon can be explained by adrenal mobilization mediated by peripheral autonomic nerves.

Measurement of the levels of cAMP and cGMP gave little information on the mode of action of opiates. Obviously, the cyclic nucleotides are so widely distributed in the nervous system (and outside of it), play such a basic role in regulation of the cellular energy metabolism, and their activity is regulated in so many ways that the attempts to relate them to the specific effects of opiates do not seem very promising. Moreover, to our present knowledge AC can be found postsynaptically due to its role in modulation of the activity of postsynaptic cells upon binding of neurotransmitters to their receptors. However, the opiate receptors are located mostly (or exclusively?) presynaptically, their main function being to modulate the neurotransmitter liberation. Consequently, these effects on the AC activity might only be secondary processes.

E. Effect on Adenosine Triphosphatase (ATPase) Activity

Contrary to AC the Mg^{++}, Ca^{++}, Na^+ or K^+ stimulated ATP-ase activities seem to be rather closely connected with the opiate receptors. These enzymes have been suggested to play regulatory role in the presynaptic storage and release of certain neurotransmitters, which is almost self-evident considering their basic role in maintenance of active transmembranal ion transport. Even the reuptake of the biogenic amines, a presynaptic process, which terminates their action is ATPase-dependent.[411] Thus, it is probably justified to ask why has the action of opiates on ATPase activity not been analyzed in detail but in one laboratory.[410,411] According to the data of Desiah and Ho[410,411] acute morphine treatment inhibited the D and NA-induced increment of Na^+-K^+-ATPase activity. In vitro morphine administration had no effect on the Na^+, K^+ and Mg^{++}-ATPase activities in mouse brain synaptosomes.[411] Naloxone alone had no effect, but reversed the in vivo exerted action of morphine.[411] How are the catecholamine sensitive Na^+- and K^+-ATPase activities inhibited by opiates? The above-mentioned authors[410,411] suggested that opiates either blocked the active sites on the enzyme molecules for the binding of catecholamines or they caused conformational changes. The synaptosomal uptake of certain neurotransmitters is believed to be dependent on Na^+, K^+-ATPase.[416] Consequently, inhibition of this enzyme results in decreased rate of uptake of catecholamines into the synaptosomes. Such an action is, however, tantamount to an initial release of D and NA followed by their depletion.

It is tempting to speculate that the previously discussed behavioral activation (Chapter 2, Volume I) observed upon administering small morphine doses is a consequence of increased levels of catecholamines due to their diminished reuptake into the presynaptic stores. Giving higher doses, a more extensive release will result in depletion of the presynaptic stores. As a result, after a short initial facilitation less catecholamine will reach the postsynaptic receptors resulting in behavioral depression in acute experiments and also in gradual development of hypersensitivity of the postsynaptic membrane upon longer exposure to the drugs.

F. Effect on the Brain Calcium and Lipids

Calcium ions play a basic role in the release of neurotransmitters[384,417] and in the membrane excitability as well. Thus any alteration of its brain concentration or its subcellular distribution may modify certain functions of the brain and those of several types of psychoactive drugs as well. As for the opiates, the ability of extracellular Ca^{++} ions to reduce the effects of morphine in vitro (Chapter 1, Volume III and Chapter 1, Volume II) is well known.

As shown in Table 14 i.c.v. administration of Ca^{++} results in attenuation of morphine analgesia.[26,329] Other bivalent cations such as Mg^{++} and Mn^{++} have a much weaker effect or do not influence the opiate actions at all (Sr^{++}, Ba^{++}, Ni^{++}, Cd^{++}, Zn^{++}).[329] Neither K^+ nor Na^+ modify the analgesia.[329] Giving Ca^{++} chelating agents such as ethylenediaminetetraacetic acid or related compounds with Ca^{++} binding potency, the analgesic effect of morphine was found potentiated.[26,329-331] The antagonist potency of Ca^{++} did not change upon tolerance development and the pA_2 value of naloxone against morphine also remained the same in Ca^{++}-treated animals.[26] These observations exclude the possibility that Ca^{++} would act directly on the opiate receptors by competing with the opiates for the latters' binding sites. Since naloxone prevents the Ca^{++} depletion upon morphine treatment,[384] this event should take place only after the opiate receptors have been activated.

Depletion of brain Ca^{++} not only potentiated the opiate analgesia, but it had an analgesic action of its own.[329,330] A close parallelism has been observed between tolerance development and replenishment of synaptosomes with Ca^{++}.[381,397,403,408,417]

The ionophore X537A greatly potentiated the opiate antagonistic action of Ca^{++} proving that these ions ought to get through the cell membrane to exert their effect.[26]

It is worth mentioning that the rare earth ion La^{+++} has also been shown to potentiate the antinociceptive action of morphine, to have analgesic effect if given alone, to act very effectively if microinjected into the PAG and its action was found antagonized by naloxone.[330] It is logical to suppose that this ion acted by competing with the Ca^{++} ions for the intracellular (synaptosomal) binding sites.

Considering that morphine and Ca^{++} are antagonists in relation of analgesia, the depletion of the latter upon administration of the former is probably not an unexpected finding. First Ross et al.[384] and Cardenas and Ross[388] showed that morphine, levorphanol, but not dextrorphan induced Ca^{++} depletion in all brain areas examined. This effect was dose-dependent and reversible by naloxone.[388] The time-course of the biochemical and behavioral effects were parallel proving the causal correlation of the two events.[388] The same group[393] and also others[297,408] demonstrated that opiates specifically depleted the synaptosomal Ca^{++}. This effect is selective in many aspects; the decrement of synaptosomal Ca^{++} content is not accompanied by alteration of the Ca^{++} content in other subcellular fractions[403] and the same treatment does not change the Na^+, K^+, and Mg^{++} contents of the synaptosomes.[411,417] The Ca^{++} depletion appears just in that subcellular fraction, which also contains the opiate receptors.[393]

The question arises: how do the opiates deplete the synaptosomal Ca^{++}? Theoretically decreased synaptosomal uptake or reduced intrasynaptosomal binding and facilitated efflux may explain this action. Yamamoto et al.[403] showed that morphine actually increased the Ca^{++} binding to certain subfractions of synaptic plasma membranes. However, the same treatment did not modify its binding to other subfractions of the synaptic plasma membrane or the synaptic vesicle fraction.[403] Consequently, simultaneously with the depletion of synaptic Ca^{++} a certain intracellular redistribution may take place. If the synaptosomal Ca^{++} binding capacity is not reduced upon morphine treatment, another mechanism must be responsible for its depletion. Moreover, acute opiate treatment in vivo or in vitro did not increase the Ca^{++} efflux from the synaptosomes, though an enhancement of this process was observed upon chronic treatment.[407] Consequently, only the third theoretical possibility remains according to the Ca^{++} uptake into the synaptosomes is decreased upon opiate treatment. And really Guerrero-Munoz et al.[408] reported that the uptake of Ca^{++} into mouse brain synaptosomes was inhibited upon morphine administration both in vivo or in vitro. Naloxone treatment reversed the morphine-induced inhibition of Ca^{++} uptake into the synaptosomal fraction and also this effect was attenuated upon tolerance development.[408]

If accepting that the depletion of synaptosomal Ca^{++} is due to a decrease in its uptake, the next question is, why is the active inward transport decelerated?

Changes in lipid metabolism might give a partial explanation. Dissociation of Ca^{++} from negatively charged binding sites is an essential step in membrane excitation. The structures, which reversibly bind the Ca^{++} ions are not known yet but certain lipids might serve as such sites. Acidic lipids such as phosphatidylserine, phosphoinositides, and gangliosides, all have high affinity for divalent cations. Accordingly, the inhibition by opiates of Ca^{++} binding to these lipids could explain the above discussed effects.

First Mulé[382] demonstrated that morphine stimulated the incorporation of labeled phosphorus into various phospholipids of guinea pig brain in vivo and in vitro as well. Upon tolerance development, this effect also was attenuated.[382] The same author[418] demonstrated the inhibition by opiates of phospolipid facilitated Ca^{++} transport in vitro. Comparing opiates of very different structures in this model experiment, a good correlation was found between the relative potencies to inhibit the Ca^{++} transport and their known analgesic potencies.[418] Nevertheless, in this system not only the narcotic agonists were found active, but also naloxone and dextrorphan.[418] But in a similar in vitro experiment the morphine-induced inhibition of Ca^{++} binding to bovine gangliosides could be antagonized by nalorphine.[419] And in a recent study[412] morphine was found to enhance the incorporation of phosphorus into phosphatidylinositol and 1-phosphatidylinositol-3,4-biphosphate but not into 1-phosphatidylinositol-4-phosphate in discrete subcellular fractions of the rat midbrain and these actions of morphine were blocked by naloxone.[412] All these data suggest that the actions of opiates are closely related to specific changes in the turnover of neural phospholipids and these changes may be responsible for their Ca^{++} depleting effect. The supposed presence of phospholipids in the structure of opiate receptors (Chapter 1, Volume III) obtains special significance in the light of these findings.

G. Other Metabolic Effects

The inhibition by opiates of *RNA synthesis* in several in vitro systems is fairly established.[387] Similar data have been obtained in vivo. Acute administration of high doses of morphine inhibited the incorporation of (^{3}H)-uridine into the mouse brain RNA, however, the same process was unaltered in the liver.[387] Moreover, morphine did not change the brain RNA concentration even upon chronic treatment.[387] Similar to morphine, methadone was also reported to inhibit the uridine incorporation.[401]

Opiates also inhibit the *protein synthesis.* Reduced rate of tyrosine incorporation was observed in rats and in mice as well, the latter effect being naloxone sensitive.[395,401]

Unfortunately, the data obtained on the effects of opiates on brain RNA and protein synthesis are difficult to interpret. A certain unspecific decrement in protein and/or RNA synthesis may probably be related to their general CNS depressant character. Considering, however, the data on the attenuation of tolerance development upon inhibition of protein synthesis (Chapter 9, Volume III), an increase in brain metabolism should also be anticipated, since the thiamin incorporation into some specific brain regions has also been found accelerated.[396] Probably the facilitated synthesis of certain proteins due to tolerance development is obscured by the general metabolic depressant effect of opiates.

Though the *histamine* liberating effect of opiates has long been known, their action on its brain level is a controversial issue. Recently one group[392] found no alteration in endogenous histamine levels in different brain areas upon acute morphine treatment, but another[413] reported on a biphasic effect, i.e., on an initial increase followed by decrement.

As for the brain *GABA* level, in one laboratory its brain concentration was not found modified[402] but others[197] found it elevated in different thalamic nuclei and in the dorsal horn upon acute administration.

In one study morphine was reported to potentiate the prostaglandin biosynthesis in the rat placenta.[406]

As for the *hyperglycemic* action of opiates, it is a very old clinical observation. This effect can be attributed to the stressor and adrenal mobilizing action of these drugs. However, this response also could be elicited by microinjecting minute amounts into the brain.[383] Of course, a pituitary-adrenal mobilization can be elicited also centrally, but the central effect excludes the possibility of a primary action mediated by the peripheral opiate receptors. Otherwise, this central hyperglycemic effect could be elicited mainly from the PAG and from other sites located close to the floor of the fourth ventricle.[383] The morphine-induced hyperglycemia could be inhibited by hydergine, but not by proranolol[420] showing the specific involvement of α-adrenergic receptors.

The altered *liver activity* seems to be centrally mediated. Morphine increases the serum glutamate-oxalate transaminase (SGOT), serum glutamate-pyruvate transaminase (SGPT),[415] and tyrosine aminotransferase (TAT)[404] activities. In the liver the gluconeogenesis and ureogenesis were found facilitated upon acute morphine treatment.[405] The former effects are naloxone sensitive.[415] Tolerance develops to all these effects upon repeated morphine administration.[404,405,415] Very important findings since abnormalities in the liver function are very common in narcotic addicts, but they are generally attributed to exogenous infections (e.g., upon repeated injections) and not to the direct opiate action. The morphine-induced activation can be elicited also by i.c.v. morphine administration,[415] showing the central origin of the effect. On the other hand, hypophysectomy prevented completely and adrenalectomy diminished the response.[415] Thus, this response might be connected with opiate receptor mediated activation of the pituitary-adrenal axis.

Finally, the action of opiates on *human serum esterase* (HSE) should be mentioned. Inhibition of HSE activity is not an opiate specific effect since many amines not related to the opioids were found active in this assay.[421] Nevertheless, this "opiate model" developed by Gero[421] resembles the real opiate receptors in several ways: the action of opiate agonists is attenuated by Na^+ (sodium effect, see Chapter 1, Volume III), differential sensitivity to the active and inactive enantiomers, etc.

REFERENCES

1. **Arunlakshana, O. and Schild, H. O.**, Some quantitative uses of drug antagonists, *Br. J. Pharmacol.*, 14, 48, 1959.
2. **Cox, B. M. and Weinstock, M.**, Quantitative studies of the antagonism by nalorphine of some of the actions of morphine-like analgesic drugs, *Br. J. Pharmacol.*, 23, 289, 1964.
3. **Takemori, A. E., Kupferberg, H. J., and Miller, J. V.**, Quantitative studies of the antagonism of morphine by nalorphine and naloxone, *J. Pharmacol. Exp. Ther.*, 169, 39, 1969.
4. **Takemori, A. E., Hayashi, G., and Smits, S. E.**, Studies on the quantitative antagonism of analgesics by naloxone and diprenorphine, *Eur. J. Pharmacol.*, 20, 85, 1972.
5. **Székely, J. I., Dunai-Kovács, Z., Miglécz, E., Rónai, A. Z., and Bajusz S.**, In vivo antagonism by naloxone of morphine, β-endorphin and a synthetic enkephalin analog, *J. Pharmacol. Exp. Ther.*, 207, 878, 1978.
6. **Smith, A. A., Albin, R., and Crofford, M.**, Heterogeneity of receptors for analgesia, respiration, and lenticular effect, in *Proceedings of the International Narcotic Research Club Meeting*, Kosterlitz, H. W., Ed., North-Holland, Amsterdam, 1976, 289.
7. **McGilliard, K. L., Tulanay, F. C., and Takemori, A. E.**, Antagonism by naloxone of morphine- and pentazocine-induced respiratory depression and analgesia and of morphine-induced hyperthermia, in *Proceedings of the International Narcotic Research Club Meeting*, Kosterlitz, H. W., Ed., North-Holland, Amsterdam, 1976, 281.
8. **Hayashi, G. and Takemori, A. E.**, The type of analgesic-receptor interaction involved in certain analgesic assays, *Eur. J. Pharmacol.*, 16, 63, 1971.
9. **Ankier, S. I.**, New hot plate tests to quantify antinociceptive and narcotic antagonist activities, *Eur. J. Pharmacol.*, 27, 1, 1974.
10. **Höllt, V., Dum, J., Bläsig, J., Schubert, P., and Herz, A.**, Comparison of in vivo and in vitro parameters of opiate receptor binding in naive and tolerant/dependent rodents, *Life Sci.*, 16, 1823, 1975.
11. **Yaksh, T. L. and Rudy, T. A.**, A dose ratio comparison of the interaction between morphine and cyclazocine with naloxone in rhesus monkeys on the shock titration task, *Eur. J. Pharmacol.*, 46, 83, 1977.
12. **Yaksh, T. L. and Rudy, T. A.**, Studies on the direct spinal action of narcotics in the production of analgesia in the rat, *J. Pharmacol. Exp. Ther.*, 202, 411, 1977.
13. **Foldes, F. F., Lunn, J. N., Moore, J., and Brown, J. M.**, N-allyl-noroxymorphone: a new potent narcotic antagonist, *Am. J. Med. Sci.*, 245, 23, 1963.
14. **Rónai, A. Z., Gráf, L., Székely, J. I., Dunai-Kovács, Z., and Bajusz, S.**, Differential behavior of LPH-(61-91)-peptide in different model systems: comparison of the opioid activities of LPH-(61-91)-peptide and its fragments, *FEBS Lett.*, 74, 182, 1977.
15. **Rónai, A. Z., Berzétei, I., Székely, J. I., Gráf, L., and Bajusz, S.**, Kinetics studies in isolated organs: tools to design analgesic peptides and to analyse their receptor effects, *Pharmacology*, 18, 18, 1979.
16. **Lord, J. A. H., Waterfield, A. A., Hughes, J., and Kosterlitz, H. W.**, Endogenous opioid peptides: multiple agonists and receptors, *Nature (London)*, 267, 495, 1977.
17. **Waterfield, A. A., Smokcum, R. W. J., Hughes, J., Kosterlitz, H. W., and Henderson, G.**, In vitro pharmacology of the opioid peptides, enkephalins and endorphins, *Eur. J. Pharmacol.*, 43, 107, 1977.
18. **Rónai, A. Z., Berzétei, I., and Bajusz, S.**, Differentiation between opioid peptides by naltrexone, *Eur. J. Pharmacol.*, 45, 393, 1977.
19. **Wüster, M., Schulz, R., and Herz, A.**, The direction of opioid agonists towards µ-, δ- and ε-receptors in the was deferens of the mouse and the rat, *Life Sci.*, 27, 163, 1980.
20. **Schulz, R., Faase, E., Wüster, M., and Herz, A.**, Selective receptors for β-endorphin on the rat deferens, *Life Sci.*, 24, 843, 1979.
21. **Smits, S. E. and Takemori, A. E.**, Quantitative studies on the antagonism by naloxone of some narcotic and narcotic-antagonist analgesics, *Br. J. Pharmacol.*, 39, 627, 1970.
22. **Gilbert, P. E. and Martin, W. R.**, The effects of morphine- and nalorphine-like drugs in the nondependent, morphine-dependent and cyclazocine-dependent chronic spinal dog, *J. Pharmacol. Exp. Ther.*, 198, 66, 1976.
23. **Martin, W. R., Eades, C. G., Thompson, J. A., Huppler, R. E., and Gilbert, P. E.**, The effects of morphine- and nalorphine-like drugs in the nondependent and morphine dependent chronic spinal dog, *J. Pharmacol. Exp. Ther.*, 197, 517, 1976.
24. **Tortella, F. C., Cowan, A., and Adler, M. W.**, EEG and behavioral effects of ethylketocyclazocine, morphine and cyclazocine in rats: differential sensitivities towards naloxone, *Neuropharmacology*, 19, 845, 1980.

25. Cowan, A., Tallarida, R. J., Maslow, J., and Adler, M. W., Quantitative assessment of the naloxone-ethylketocyclazocine interaction using the tail compression test, *Pharmacologist,* 19, 140, 1977.
26. Harris, R. A., Loh, H. H., and Way, E. L., Effects of divalents cations, cation chelators, and an ionophore on morphine analgesia and tolerance, *J. Pharmacol. Exp. Ther.,* 195, 488, 1975.
27. Takemori, A. E., Oka, T., and Nishiyama, N., Alteration of analgesic receptor-antagonist interaction induced by morphine, *J. Pharmacol. Exp. Ther.,* 186, 261, 1973.
28. Tulunay, F. C. and Takemori, A. E., The increased efficacy of narcotic antagonists induced by various narcotic analgesics, *J. Pharmacol. Exp. Ther.,* 190, 395, 1974.
29. Tulunay, F. C. and Takemori, A. E., Further studies on the alteration of analgesic receptor-antagonist interaction induced by morphine, *J. Pharmacol. Exp. Ther.,* 190, 401, 1974.
30. Kitano, T. and Takemori, A. E., Further studies on the enhanced affinity of opioid receptors for naloxone in morphine-dependent mice, *J. Pharmacol. Exp. Ther.,* 209, 456, 1979.
31. Wong, C.-L. and Bentley, G. A., Increased antagonist potency of naloxone caused by morphine pretreatment in mice, *Eur. J. Pharmacol.,* 47, 415, 1978.
32. Lange, D. G., Fujimoto, J. M., Roerig, S. C., and Wang, R. I. H., Morphine induced sensitization to naloxone: enhanced disposition of naloxone to the brain, *Pharmacologist,* 18, 121, 1976.
33. Manara, L., Aldinio, C., and Cerletti, C., Effect of naloxone on morphine disposition and brain levels in rats, in *Proc. 7th Int. Congr. Pharm.,* Paris, France, 1978, Abstract, 752.
34. Jasinski, D. R., Assessment of the abuse potentiality of morphine-like drugs (methods used in man), in *Drug Addiction,* Martin, W. R., Ed., Springer-Verlag, New York, 1977, Chap. 3.
35. Isbell, H., Belleville, R. E., Fraser, H. F., Wikler, A., and Logan, C. R., Studies on lysergic acid diethylamide (LSD-25). Effects in former morphine addicts and development of tolerance during chronic intoxication, *Arch. Neurol. Psychiatr. (Chicago),* 76, 468, 1956.
36. Fraser, H. F. and Isbell, H., Human pharmacology and addiction liabilities of phenazocine and levophenacylmorphan, *Bull. Narcot.,* 12, 15, 1960.
37. Hill, H. E., Haertzen, C. A., Wolbach, A. B., Jr., and Miner, E. J., The addiction research center inventory: standardization of scales which evaluate subjective effects of morphine, amphetamine, pentobarbital, alcohol, LSD-25, pyrahexyl and chlorpromazine, *Psychopharmacologia,* 4, 167, 1963.
38. Overton, D. A., State dependent or "dissociated" learning produced with pentobarbital, *J. Comp. Physiol. Psychol.,* 57, 3, 1964.
39. Hill, H. E., Jones, B. E., and Bell, E. C., State dependent control of discrimination by morphine and pentabarbital, *Psychopharmacologia,* 22, 305, 1971.
40. Colpaert, F. C. and Niemegeers, C. J. E., On the narcotic cuing action of fentanyl and other narcotic analgesic drugs, *Arch. Int. Pharmacodyn.,* 217, 170, 1975.
41. Wüster, M. and Herz, A., Opiate agonist action of antidiarrheal agents, *in vitro* and *in vivo*-findings in support for selective action, *Naunyn-Schmiedeb. Arch. Pharmacol.,* 301, 187, 1978.
42. Gianutsos, G. and Lal, H., Effect of loperamide, haloperidol and methadone in rats trained to discriminate morphine form saline, *Psychopharmacologia,* 41, 267, 1975.
43. Shannon, H. E. and Holtzman, S. G., Evaluation of the discriminative effects of morphine in the rat, *J. Pharmacol. Exp. Ther.,* 198, 54, 1976.
44. Shannon, H. E. and Holtzman, S. G., Further evaluation of the discriminative effects of morphine in the rat, *J. Pharmacol. Exp. Ther.,* 201, 55, 1977.
45. Colpaert, F. C., Niemegeers, C. J. E., and Janssen, P. A. J., Discriminative stimulus properties of analgesic drugs: narcotic versus non-narcotic analgesics, *Arch. Int. Pharmacodyn.,* 220, 329, 1976.
46. Colpaert, F. C., Niemegeers, C. J. E., and Janssen, P. A. J., On the ability of narcotic antagonists to produce the narcotic-cue, *J. Pharmacol. Exp. Ther.,* 197, 180, 1976.
47. Hirschhorn, I. D. and Rosecrans, J. A., Generalization of morphine and lysergic acid diethylamide (LSD) stimulus properties to narcotic analgesics, *Psychopharmacology,* 47, 65, 1976.
48. Hirschhorn, I. D., Pentazocine, cyclazocine, and nalorphine as discriminative stimuli, *Psychopharmacology,* 54, 289, 1977.
49. Teal, J. J. and Holtzman, S. G., Discriminative stimulus effects of cyclazocine in the rat, *J. Pharmacol. Exp. Ther.,* 212, 368, 1980.
50. Haertzen, C. A., Subjective effects of narcotic antagonists cyclazocine and nalorphine on the Addiction Research Center Inventory (ARCI), *Psychopharmacologia,* 18, 366, 1970.
51. Lal, H., Narcotic dependence, narcotic action, and dopamine receptors, *Life Sci.,* 17, 483, 1975.
52. Colpaert, F. C., Niemegeers, C. J. E., and Janssen, P. A. J., Fentanyl and apomorphine: assymetrical generalization of discriminative stimulus properties, *Neuropharmacology,* 15, 541, 1976.
53. Miksic, S., Shearman, G., and Lal, H., Generalization study with some narcotic and nonnarcotic drugs in rats trained for morphine-saline discrimination, *Psychopharmacology,* 60, 103, 1978.
54. Gianutsos, G. and Lal, H., Selective interaction of drugs with a discriminable stimulus associated with narcotic action, *Life Sci.,* 19, 91, 1976.
55. Golembiowska, N. K., Pilc, A., and Vetulani, J., Opiates and specific receptor binding of (^{3}H) - clonidine, *J. Pharm. Pharmacol.,* 32, 70, 1980.

56. Paalzow, L., Analgesia produced by clonidine in mice and rats, *J. Pharm. Pharmacol.*, 26, 361, 1974.
57. Paalzow, G. and Paalzow, L., Clonidine antiociceptive activity: Effects of drugs influencing central monoaminergic and cholinergic mechanisms in the rat, *Naunyn-Schmiedeb. Arch. Pharmacol.*, 292, 119, 1976.
58. Colpaert, F. C., Niemegeers, C. J. E., and Janssen, P. A. J., Differential haloperidol effect on two indices of fentanyl-saline discrimination, *Psychopharmacology*, 53, 169, 1977.
59. Schechter, M. D., Lack of blockade of central dopaminergic receptors by narcotics: comparison with chlorpromazine, *Eur. J. Pharmacol.*, 49, 279, 1978.
60. Rosecrans, J. A., Goodloe, M. H., Jr., and Bennett, G. J., Morphine as a discriminative cue: effects of amine depletors and naloxone, *Eur. J. Pharmacol.*, 21, 252, 1973.
61. Winter, J. C., Morphine and ethanol as discriminative stimuli: absence of antagonism by p-chlorophenylalanine methyl ester, cinanserin, or BC-105, *Psychopharmacology*, 53, 159, 1977.
62. Shannon, H. E. and Holtzman, S. G., Discriminative effects of morphine administered intracerebrally in the rat, *Life Sci.*, 21, 585, 1977.
63. Colpaert, F. C., Niemegeers, C. J. E., Janssen, P. A. J., and van Ree, J. M., Narcotic cueing properties of intraventricularly administered sufentanil, fentanyl, morphine and Met-enkephalin, *Eur. J. Pharmacol.*, 47, 115, 1978.
64. Rosecrans, J. A. and Krynock, G. M., A possible role of the PAG in the mediation of subjective effects of morphine, *Pharmacologist*, 19, 171, 1977.
65. Van Ree, J. M., Smyth, D. G., and Colpaert, F. G., Dependence creating properties of lipotropin C-fragment (β-endorphin): evidence for its internal control of behavior, *Life Sci.*, 24, 495, 1979.
66. D'Amour, F. E. and Smith, D. L., A method for determining loss of pain sensation, *J. Pharmacol.*, 72, 74, 1941.
67. Ben-Bassat, J., Peretz, E., and Sulman, F. G., Analgesimetry and ranking of analgesic drugs by the receptacle method, *Arch. Int. Pharmadocyn.*, 122, 434, 1959.
68. Sewell, R. D. E. and Spencer, P. S. J., Antinociceptive activity of narcotic agonist and partial agonist analgesics and other agents in the tail-immersion test in mice and rats, *Neuropharmacology*, 15, 683, 1976.
69. Stewart, J. M., Getto, C. J., Nelder, K., Reeve, E. B., Krivoy, W. A., and Zimmermann, E., Substance P and analgesia, *Nature (London)*, 262, 784, 1976.
70. Winter, C. A. and Flataker, L., The effect of cortisone, desoxycorticosterone, and adrenocorticotrophic hormone upon the responses of animals to analgesic drugs, *J. Pharmacol. Exp. Ther.*, 103, 93, 1951.
71. Haffner, F., Experimentelle Prüfung schmerzstillender Mittel, *Dtsch. Med. Wschr.*, 55, 731, 1929.
72. Parkes, M. W., An examination of central actions characteristic of scopolamine: comparison of central and peripheral activity in scopolamine, atropine and some synthetic basic esters, *Psychopharmacologia*, 7, 1, 1965.
73. Grewal, R. S., A method for testing analgesics in mice, *Br. J. Pharmacol.*, 7, 433, 1952.
74. Paalzow, L. and Paalzow, G., Studies on the relationship between morphine analgesia and the brain catecholamines in mice, *Acta Pharmacol. Toxicol.*, 30, 104, 1971.
75. Dahlström, B., Paalzow, G., and Paalzow, L., A pharmacokinetic approach to morphine analgesia and its relation to regional turnover of rat brain catecholamines, *Life Sci.*, 17, 11, 1975.
76. Evans, W. O., A new technique for the investigation of some analgesic drugs on a reflexive behavior in the rat, *Psychopharmacologia*, 2, 318, 1961.
77. Pert, A., The cholinergic system and nociception in the primate: interactions with morphine, *Psychoparmacologia*, 44, 131, 1975.
78. Weiss, B. and Laties, V. G., The psychophysics of pain and analgesia in animals, in *Animal Psychophysics. The Design and Conduct of Sensory Experiments*, Stebbins, W. C., Ed., Appleton-Century-Crofts, New York, 1970, 185.
79. Eddy, N. B., Fuhrmeister Touchberry, C., and Lieberman, J. E., Synthetic analgesics I. Methadone isomers and derivatives, *J. Pharmacol.*, 98, 121, 1950.
80. Siegmund, E., Cadmus, R., and Lu, G., A method for evaluating both non-narcotic and narcotic analgesics, *Proc. Soc. Exp. Biol. Med.*, 95, 729, 1957.
81. Hendershot, L. C. and Forsaith, J., Antagonism of the frequency of phenylquinone-induced writhing in the mouse by weak analgesics and nonanalgesics, *J. Pharmacol. Exp. Ther.*, 125, 237, 1959.
82. Witkin, L. B., Heubner, C. F., Galdi, F., O'Keefe, E., Spitaletta, P., and Plummer, A. J., Pharmacology of 2-amino-indane hydrochloride (SU-8629): a potent non-narcotic analgesic, *J. Pharmacol. Exp. Ther.*, 133, 400, 1961.
83. Chernov, H. I., Wilson, D. E., Fowler, F., and Plummer, A. J., Non-specificity of the mouse writhing test, *Arch. Int. Pharmacodyn.*, 167, 171, 1967.
84. Randall, L. O. and Selitto, J. J., A method for measurement of analgesic activity of inflamed tissue, *Arch. Int. Pharmacodyn.*, 111, 409, 1957.

85. Pearl, J., Aceto, M. D., and Fitzgerald, J. J., Differences in antiwrithing activity of morphine and nalorphine over time and in scopes of the dose response lines, *Psychopharmacologia*, 13, 341, 1968.
86. Irwin, S., Houde, R. W., Bennett, D. R., Hendershot, L. C., and Seevers, M. H., The effects of morphine, methadone and meperidine on some reflex responses of spinal animals to nociceptive stimulation, *J. Pharmacol. Exp. Ther.*, 101, 132, 1950.
87. Dewey, W. L., Snyder, J. W., Harris, L. S., and Howes, J. F., The effect of narcotics and narcotic antagonists on the tail flick response in spinal mice, *J. Pharm. Pharmacol.*, 21, 548, 1969.
88. Wilcox, G. L. and Dewey, W. L., Morphine analgesia in spinal mice and rats: cholinergic mechanisms, *Pharmacologist*, 18, 213, 1976.
89. Fu, T. C. and Dewey, W. L., Morphine antinociception: evidence for the release of endogenous substance(s), *Life Sci.*, 25, 53, 1979.
90. Satoh, M. and Takagi, H., Enhancement by morphine of the central descending inhibitory influence on spinal sensory transmission, *Eur. J. Pharmacol.*, 14, 60, 1971.
91. Le Bars, D., Rivot, J. P., Guilbaud, G., Menetrey, D., and Besson, J. M., The depressive effect of morphine on the C fibre response of dorsal horn neurones in tbe spinal rat pretreated or not by pCPA, *Brain Res.*, 176, 337, 1979.
92. Fields, H. L., Basbaum, A. I., Clanton, C. H., and Anderson, S. D., Nucleus raphe magnus inhibition of spinal cord dorsal horn neurones, *Brain Res.*, 126, 441, 1977.
93. Willis, W. D., Haber, L. H., and Martin, R. F., Inhibition of spinothalamic tract cells and interneurones by brain stem stimulation in the monkey, *J. Neurophysiol.*, 40, 968, 1977.
94. Samanin, R., Gumulka, W., and Valzelli, L., Reduced effect of morphine in midbrain raphe lesioned rats, *Eur. J. Pharmacol.*, 10, 339, 1970.
95. Garau, L., Mulas, M. L., and Pepeu, G., The influence of raphe lesions on the effect or morphine on nociception and cortical ACh output, *Neuropharmacology*, 14, 259, 1975.
96. Mayer, D. J. and Price, D. D., Central nervous system mechanisms of analgesia, *Pain*, 2, 379, 1976.
97. Sasa, M., Munekiyo, K., Osumi, Y., and Takaori, S., Attenuation of morphine analgesia in rats with lesions of the locus coeruleus and dorsal raphe nucleus, *Eur. J. Pharmacol.*, 42, 53, 1977.
98. Deakin, J. F. W., Dostrovsky, J. O., and Longden, A., The role of periaqueductal grey matter and of spinal serotonergic pathway in morphine analgesia, *J. Physiol. (London)*, 275, 67, 1978.
99. Chance, W. T., Krynock, G. M., and Rosecrans, J. A., Effects of medial raphe and raphe magnus lesions on the analgesic activity of morphine and methadone, *Psychopharmacology*, 56, 133, 1978.
100. Miranda, F., Candelaresi, G., and Samanin, R., Analgesic effect of etorphine in rats with selective depletions of brain monoamines, *Psychopharmacology*, 58, 105, 1978.
101. Genovese, E., Zonta, N., and Mantagezza, P., Decreased antinociceptive activity of morphine in rats pretreated intraventricularly with 5,6-dihydroxytryptamine, a long-lasting selective depletor of brain serotonin, *Psychopharmacologia*, 32, 359, 1973.
102. Snelgar, R. and Vogt, M., The effect of morphine on the turnover of 5-hydroxytryptamine in discrete parts of the rat brain, *J. Physiol. (London)*, 284, 128, 1978.
103. Bläsig, J., Reinhold, K., and Herz, A., Effect of 6-hydroxydopamine, 5,6-dihydroxytryptamine and raphe lesions on the antinociceptive actions of morphine in rats, *Psychopharmacologia*, 31, 111, 1973.
104. Pottoff, P., Valentino, D., and Lal, H., Attenuation of morphine analgesia by lesions of the preoptic forebrain region in the rat, *Life Sci.*, 24, 421, 1979.
105. Kostowski, W., Jerlicz, M., Bidzinski, A., and Hauptmann, M., Studies on the effect of lesions of the ventral noradrenergic tract on the antinociceptive action of morphine, *Psychopharmacology*, 57, 189, 1978.
106. Hammond, D. L. and Proudfit, H. K., Potentiation of morphine analgesia by lesions of the nucleus locus coeruleus, *Pharmacologist*, 19, 140, 1977.
107. Hammond, D. L. and Proudfit, H. K., Effects of locus coeruleus lesions on morphine-induced antinociception, *Brain Res.*, 188, 79, 1980.
108. Nakamura, K., Kuntzman, R., Maggio, A. C., Augulis, V., and Conney, A. H., Influence of 6-hydroxydopamine on the effect of morphine on the tail-flick latency, *Psychopharmacologia*, 31, 177, 1973.
109. Yeung, J. C., Yaksh, T. L., and Rudy, T. A., Effects of brain lesions on the antinociceptive properties of morphine in rats, *Clin. Exp. Pharmacol. Physiol.*, 2, 261, 1975.
110. Jhamandas, K. and Sutak, M., Morphine-naloxone interaction in the central cholinergic system: the influence of subcortical lesioning and electrical stimulation, *Br. J. Pharmacol.*, 58, 101, 1976.
111. Jancsó, G. and Jancsó-Gabor, A., Effect of capsaicin on morphine analgesia-possible involvement of hypothalamic structures, *Naunyn-Schmiedeb. Arch. Pharmacol.*, 311, 285, 1980.
112. Shah, Y. and Dostrovsky, J. O., Electrophysiological evidence for a projection of the periaqueductal gray matter to nucleus raphe magnus in cat and rat, *Brain Res.*, 193, 534, 1980.
113. Babbini, M., and Davis, W. M., Time-dose relationships for locomotor activity effects of morphine after ecute or repeated treatment, *Br. J. Pharmacol.*, 46, 213, 1972.

114. Vasko, M. R. and Domino, E. F., Tolerance development to the biphasic effects of morphine on locomotor activity and brain acetylcholine in the rat, *J. Pharmacol. Exp. Ther.*, 207, 848, 1978.
115. Domino, E. F., Vasko, M. R., and Wilson, A. E., Mixed depressant and stimulant actions of morphine and their relationship to brain acetylcholine, *Life Sci.*, 18, 361, 1976.
116. Martin, W. R. and Sloan, J. W., Neuropharmacology and neurochemistry of subjective effects, analgesia, tolerance and dependence produced by narcotic analgesics, in *Drug Addiction*, Martin, W. R., Ed., Springer-Verlag, New York, 1977, Chap. 1.
117. Fog, R., Behavioral effects in rats of morphine and amphetamine and of a combination of the two drugs, *Psychopharmacologia*, 16, 305, 1970.
118. Ayhan, I. H. and Randrup, A., Behavioral and pharmacological studies on morphine-induced excitation of rats. Possible relation to brain catecholamines, *Psychopharmacologia*, 29, 317, 1973.
119. Martin, W. R., Wikler, A., Eades, C. G., and Pescor, F. T., Tolerance to and physical dependence on morphine in rats, *Psychopharmacologia*, 4, 247, 1963.
120. Oka, T. and Hosoya, E., Effects of humoral modulators and naloxone on morphine-induced changes in the spontaneous locomotor activity of the rat, *Psychopharmacology*, 47, 243, 1976.
121. Pert, A. and Sivit, C., Neuroanatomical focus for morphine and enkephalin-induced hypermotility, *Nature (London)*, 265, 645, 1977.
122. Rethy, C. R., Smith, C. B., and Villareal, J. E., Effects of narcotic analgesics upon the locomotor activity and brain catecholamine content of the mouse, *J. Pharmacol. Exp. Ther.*, 176, 472, 1971.
123. Jacquet, Y. F. and Lajtha, A., Paradoxical effects after microinjection of morphine in the periaqueductal gray matter in the rat, *Science*, 185, 1055, 1974.
124. Jacquet, Y. F., Opiate effects after adrenocorticotropin or β-endorphin injection in the preiaqueductal gray matter of rats, *Science*, 201, 1032, 1978.
125. Labella, F. S., Pinsky, C., and Havlicek, V., Morphine derivatives with diminished opiate receptor potency show enhanced central excitatory activity, *Brain Res.*, 174, 263, 1979.
126. Bergmann, F., Chaimovitz, M., and Pasternak, V., Dual action of morphine and related drugs on compulsive gnawing of rats, *Psychopharmacologia*, 46, 87, 1976.
127. Khalsa, J. H. and Davis, W. M., Motility response to morphine and amphetamine during chronic inhibition of tyrosine hydroxylase or dopamine β-hydroxylase, *J. Pharmacol. Exp. Ther.*, 202, 182, 1977.
128. Jacquet, Y. F. and Lajtha, A., Morphine action at central nervous system sites in rat: analgesia or hyperalgesia depending on site and dose, *Science*, 182, 490, 1973.
129. Dhasmana, K. M., Dixit, K. S., Jaju, B. P., and Gupta, M. L., Role of central dopaminergic receptors in manic response of cats to morphine, *Psychopharmacologia*, 24, 380, 1972.
130. Sharkawi, M. and Goldstein, A., Antagonism by physostigmine of the "running fit" caused by levorphanol, a morphine congener, in mice, *Br. J. Pharmacol.*, 37, 123, 1969.
131. Goldstein, A. and Sheehan, P., Tolerance to opioid narcotics I. Tolerance to the "running fit" caused by levorphanol in the mouse, *J. Pharmacol. Exp. Ther.*, 169, 175, 1969.
132. Catlin, D. H., George, R., and Li, C. H., β-endorphin: pharmacologic and behavioral activity in cats after low intravenous doses, *Life Sci.*, 23, 2147, 1978.
133. Székely, J. I., Dunai-Kovács, Z., Miglécz, E., and Tarnawa, I., The *in vivo* pharmacology of (D-Met^2, Pro^5)-enkephalinamide, in *Endorphins '78*, Gráf, L., Palkovits, M., and Rónai, A. Z., Eds., Publishing House of the Hungarian Acad. Sci., Budapest, Hungary, 1978, 319.
134. Cools, A. R., Janssen, H. J., and Broekkamp, C. L. E., The differential role of caudate nucleus and linear raphe nucleus in the initiation and the maintenance of morphine-induced behavior in cats, *Arch. Int. Pharmacodyn.*, 210, 163, 1974.
135. Hollinger, M., Effect of reserpine, α-methyl-*p*-tyrosine, *p*-chlorophenylalanine and pargyline on levorphanol-induced running activity in mice, *Arch. Int. Pharmacodyn.*, 179, 419, 1969.
136. Oliverio, A. and Castellano, C., Genotype-dependent sensitivity and tolerance to morphine and heroin: dissociation between opiate-induced running and analgesia in the mouse, *Psychopharmacologia*, 39, 13, 1974.
137. Shuster, L., Hannam, R. V., and Boyle, W. E., Jr., A simple method for producing tolerance to dihydromorphinone in mice, *J. Pharmacol. Exp. Ther.*, 140, 149, 1963.
138. Hecht, A. and Schiørring, E., Behavioral effects of low and high acute doses of morphine in solitary mice, *Psychopharmacology*, 64, 73, 1979.
139. Székely, J. I., Miglécz, E., and Rónai, A. Z., Biphasic effects of a potent enkephalin analog (D-Met^2, Pro^5)-enkephalinamide and morphine on locomotor activity in mice, *Psychopharmacology*, 71, 299, 1981.
140. Kuschinsky, K. and Hornykiewicz, O., Effects of morphine on striatal dopamine metabolism: possible mechanism of its opposite effect on motor activity in rats and in mice, *Eur. J. Pharmacol.*, 26, 41, 1974.
141. Judson, B. A. and Goldstein, A., Genetic control of opiate-induced locomotor activity in mice, *J. Pharmacol. Exp. Ther.*, 206, 56, 1978.

142. Tepper, P. and Woods, J. H., Changes in locomotor activity and naloxone-induced jumping in mice produced by WIN 35, 197-2 (ethylketazocine) and morphine, *Psychopharmacology*, 58, 125, 1978.
143. Beecham, I. J. and Handley, S. L., Potentiation of catalepsy induced by narcotic agents during Haffner's test for analgesia, *Psychopharmacologia*, 40, 157, 1974.
144. Costall, B. and Naylor, R. J., Neuroleptic and non-neuroleptic catalepsy, *Arzneim. Forsch.*, 23, 674, 1973.
145. Morpurgo, C., Effects of antiparkinson drugs on a phenothiazine-induced catatonic reaction, *Arch. Int. Pharmacodyn.*, 137, 84, 1962.
146. Costall, B. and Naylor, R. J., On catalepsy and catatonia and the predictability of the catalepsy test for neuroleptic activity, *Psychopharmacologia*, 34, 233, 1974.
147. Browne, R. G., Derrington, D. C., and Segal, D. S., Comparison of opiate- and opioid-peptide-induced immobility, *Life Sci.*, 24, 933, 1979.
148. Ezrin, W. C., Muller, P., and Seeman, P., Catalepsy induced by morphine or haloperidol: effects of apomorphine and anticholinergic drugs, *Can. J. Physiol. Pharmacol.*, 54, 516, 1976.
149. Kuschinsky, K. and Hornykiewicz, O., Morphine catalepsy in the rat: relation to striatal dopamine metabolism, *Eur. J. Pharmacol.*, 19, 119, 1972.
150. Costall, B. and Naylor, R. J., Serotonergic involvement with the stereotypy/catalepsy induced by morphine-like agents in the rat, *J. Pharm. Pharmacol.*, 27, 67, 1975.
151. Randrup, A. and Munkvad, I., Stereotyped activities produced by amphetamine in several animal species and man, *Psychopharmacologia*, 11, 300, 1967.
152. Ernst, A. M., Relation between the action of dopamine and apomorphine and their O-methylated derivatives upon the CNS, *Psychopharmacologia*, 7, 391, 1965.
153. Scheel-Krüger, J., Golembiowska, K., and Moglinicka, E., Evidence for increased apomorphine-sensitive dopaminergic effects after acute treatment with morphine, *Psychopharmacology*, 53, 55, 1977.
154. Vedernikov, Yu. P. and Afrikanov, I. I., On the role of a central adrenergic mechanism in morphine analgesia, *J. Pharm. Pharmacol.*, 21, 845, 1969.
155. Vedernikov, Yu. P., Interaction of amphetamine, apomorphine, disulfiram with morphine and the role played by catecholamines in morphine, analgesic action, *Arch. Int. Pharmacodyn.*, 182, 59, 1969.
156. Bergmann, F., Chaimovitz, M., Pasternak, V., and Ramu, A., Compulsive gnawing in rats after implantation of drugs into the ventral thalamus, a contribution to the mechanism of morphine action, *Br. J. Pharmacol.*, 51, 197, 1974.
157. Iwamoto, E. T. and Way, E. L., Circling behavior and stereotypy induced by intranigral opiate microinjections, *J. Pharmacol. Exp. Ther.*, 203, 347, 1977.
158. Iwamato, E. T. and Way, E. L., Stereotyped behavior induced by intranigral opiate microinjection in rats, in *Proc. Int. Narcotic Res. Conf.*, Van Ree, J. M., and Terenius, L., Eds., Elsevier/North-Holland, Amsterdam, 1978, 427.
159. Verri, R. A., Graeff, F. G., and Corrado, A. P., Antagonism of morphine analgesia by reserpine and α-methyltyrosine and the role played by catecholamines in morphine analgesic action, *J. Pharm. Pharmacol.*, 19, 264, 1967.
160. Takagi, H., Takashima, T., and Kimura, K., Antagonism of the analgesic effect of morphine in mice by tetrabenazine and reserpine, *Arch. Int. Pharmacodyn.*, 149, 484, 1964.
161. Major, C. T. and Pleuvry, B. J., Effects of α-methyl-p-tyrosine, p-chlorophenylalanine, L-β-/3,4-dihydroxyphenyl/alanine, 5-hydroxytryptophan and diethyldithiocarbamate on the analgesic activity of morphine and methylamphetamine in the mouse, *Br. J. Pharmacol.*, 42, 512, 1971.
162. Schaumann, W., Beeinflussung der analgetischen Wirkung des Morphins durch Reserpin, *Naunyn-Schmiedeb. Arch. Pharmakol.*, 235, 1, 1958.
163. Harris, L. S., Dewey, W. L., Howes, J. F., Kennedy, J. S., and Pars, H., Narcotic-antagonist analgesics: interactions with cholinergic systems, *J. Pharmacol. Exp. Ther.*, 169, 17, 1969.
164. Harris, L. S., Central neurohumoral systems involved with narcotic agonists and antagonists, *Fed. Proc. Fed. Am. Soc. Exp. Biol.*, 29, 28, 1970.
165. Grossmann, W., Jurna, I., Nell, T., and Theres, C., The dependence of the antinociceptive effect of morphine and other analgesic agents on spinal motor activity after central monoamine depletion, *Eur. J. Pharmacol.*, 24, 67, 1973.
166. Garcia Leme, J. and Rochae Silva, M., Analgesic action of chlorpromazine and reserpine in relation to that of morphine, *J. Pharm. Pharmacol.*, 13, 734, 1961.
167. Contreras, E., Quijada, L., and Tamayo, L., A comparative study of the effects of reserpine and p-chlorophenylalanine on morphine analgesia in mice, *Psychopharmacologia*, 28, 319, 1973.
168. Tripod, J. and Gross, F., Unterschiedliche Beeinflussung der analgetischen und der erregenden Wirkung von Morphin durch zentral dapmfende Pharmaka, *Helv. Physiol. Pharmacol. Acta*, 15, 105, 1957.

169. Ross, J. W. and Ashford, A., The effect of reserpine and α-methyldopa on the analgesic action of morphine in the mouse, *J. Pharm. Pharmacol.*, 19, 709, 1967.
170. Dandiya, P. C. and Menon, M. K., Studies on central nervous system depressants. III. Influence of some tranquillizing agents on morphine analgesia, *Arch. Int. Pharmacodyn.*, 141, 223, 1963.
171. Tardos, L. und Jobbágyi, Z., Wirkung von Reserpin auf den Effekt der Analgetika, *Acta Physiol. Acad. Sci. Hung.*, 13, 171, 1958.
172. Schneider, J. A., Reserpine antagonism of morphine analgesia in mice, *Proc. Soc. Exp. Biol. Med.*, 87, 614, 1954.
173. Sigg, E. B., Caprio, G., and Schneider, J. A., Synergism of amines and antagonism of reserpine to morphine analgesia, *Proc. Soc. Exp. Biol. Med.*, 97, 97, 1958.
174. Medakovič, M. and Banič, B., The action of reserpine and α-methyl-p-tyrosine on the analgesic effect of morphine in rats and mice, *J. Pharm. Pharmacol.*, 16, 198, 1964.
175. Rudzik, A. D. and Mennear, J. H., Antagonism of analgesics by amine-depleting agents, *J. Pharm. Pharmacol.*, 17, 326, 1965.
176. Verri, R. A., Graeff, F. G., and Corrado, A. P., Effect of reserpine and alpha-methyl-tyrosine on morphine analgesia, *Int. J. Neuropharmacol.*, 7, 283, 1968.
177. Dewey, W. L., Harris, L. S., Howes, J. F., and Nuite, J. A., The effect of various neurohumoral modulators on the activity of morphine and the narcotic antagonists in the tail flick and phenylquinone tests, *J. Pharmacol. Exp. Ther.*, 175, 435, 1970.
178. Fennessy, M. R. and Lee, J. R., Modification of morphine analgesia by drugs affecting adrenergic and tryptaminergic mechanisms, *J. Pharm. Pharmacol.*, 22, 930, 1970.
179. Sparkes, C. G. and Spencer, P. S. J., Antinociceptive activity of morphine after injection of biogenic amines in the cerebral ventricles of the conscious rat, *Br. J. Pharmacol.*, 42, 230, 1971.
180. Botting, R., Bower, S., Eason, C. T., Hutson, P. H., and Wells, L., Modification by monoamine oxidase inhibitors of the analgesic, hypothermic and toxic action of morphine and pethidine in mice, *J. Pharm. Pharmacol.*, 30, 36, 1978.
181. Fuentes, J. A., Garzon, J., and Del Rio, J., Potentiation of morphine analgesia in mice after inhibition of brain type B monoamine oxidase, *Neuropharmacology*, 16, 857, 1977.
182. Koe, K. B. and Weissman, A., p-chlorophenylalanine: a specific depletor of brain serotonin, *J. Pharmacol. Exp. Ther.*, 154, 499, 1966.
183. Tenen, S. S., Antagonism of the analgesic effect of morphine and other drugs by p-chlorophenylalanine, a serotonin depletor, *Psychopharmacologia*, 12, 278, 1968.
184. Takemori, A. E., Tulunay, F. C., and Yano, I., Differential effects on morphine analgesia and naloxone antagonism by biogenic amine modifiers, *Life Sci.*, 17, 21, 1975.
185. Görlitz, B.-D. and Frey, H.-H., Central monoamines and antinociceptive drug action, *Eur. J. Pharmacol.*, 20, 171, 1972.
186. Tilson, H. A. and Rech, R. H., The effects of p-chlorophenylalanine on morphine analgesia, tolerance and dependence development in two strains of rats, *Psychopharmacologia*, 35, 45, 1974.
187. Tulunay, F. C., Yano, I., and Takemori, A. E., The effect of biogenic amine modifers on morphine analgesia and its antagonism by naloxone, *Eur. J. Pharmacol.*, 35, 285, 1976.
188. Lee, R. L., Sewell, R. D. E., and Spencer, P. S. J., Importance of 5-hydroxytryptamine in the antinociceptive activity of the leucine-enkephalin derivative, D-Ala^2, Leu^5-enkephalin (BW 180C), in the rat, *Eur. J. Pharmacol.*, 47, 251, 1978.
189. Sewell, R. D. E. and Spencer, P. S. J., Biogenic amines and the antinociceptive activity of agents with a non-opiate structure, *J. Pharm. Pharmacol.*, 26, 92P, 1974.
190. Calcutt, C. R. and Spencer, P. S. J., Activities of narcotic-antagonist analgesics following the intraventicular injection of various substances, *Br. J. Pharmacol.*, 41, 401, 1971.
191. Calcutt, C. R., Doggett, N. S., and Spencer, P. S. J., Modification of the anti-nociceptive activity of morphine by centrally administered ouabain and dopamine, *Psychopharmacologia*, 21, 111, 1971.
192. Pleuvry, B. J. and Tobias, M. A., Comparison of the antinociceptive activities of physostigmine, oxotremorine and morphine in the mouse, *Br. J. Pharmacol.*, 43, 706, 1971.
193. Nakamura, K., Kuntzman, R., Maggio, A., and Conney, A. H., Decrease in morphine's analgesic action and increase in its cataleptic action by 6-hydroxydopamine injected bilaterally into putamen and caudate areas: partial restoration by l-dopa plus decarboxylase inhibition, *Neuropharmacology*, 12, 1153, 1973.
194. Mudgill, L., Friedhoff, A. J., and Tobey, J., Effect of intraventricular administrations of epinephrine, norepinephrine, dopamine, acetylcholine and physostigmine on morphine-analgesia in mice, *Arch. Int. Pharmacodyn.*, 210, 85, 1974.
195. Liu, S.-J. and Wang, R. I. H., Increased analgesia and alterations in distribution and metabolism of methadone by desipramine in the rat, *J. Pharmacol. Exp. Ther.*, 195, 94, 1975.
196. Yoneda, Y., Takashima, S., and Kuriyama, K., Possible involvement of GABA in morphine analgesia, *Biochem. Pharmacol.*, 25, 2669, 1976.

197. Kuriyama, K. and Yoneda, Y., Alterations in distribution and metabolism of γ-aminobutyric acid (GABA) in the central nervous system following morphine administration, *Jpn. J. Pharmacol.*, 26, 18, 1976.
198. Sugrue, M. F. and McIndewar, I., Effect of blockade of 5-hydroxytryptamine reuptake on drug induced antinociception in the rat, *J. Pharm. Pharmacol.*, 28, 447, 1976.
199. Tofanetti, O., Albiero, L., Galatulas, I., and Genovese, E., Enhancement of propoxyphene- induced analgesia by doxepin, *Psychopharmacology*, 51, 213, 1977.
200. Mantegazza, P., Tammiso, R., Vicentini, L., Zambotti, F., and Zonta, N., Muscimol antagonism of morphine analgesia in rats, *Br. J. Pharmacol.*, 67, 103, 1979.
201. Malseed, R. T. and Goldstein, F. J., Enhancement of morphine analgesia by tricyclic antidepressants, *Neuropharmacology*, 18, 827, 1979.
202. Gonzalez, J. P., Sewell, R. D. E., and Spencer, P. S. J., Antinociceptive activity of opiates in the presence of the antidepressant agent nomifensine, *Neuropharmacology*, 19, 613, 1980.
203. Malec, D. and Langwinski, R., Effect of quipazine and fluoxetine on analgesic-induced catalepsy and antinociception in the rat, *J. Pharm. Pharmacol.*, 32, 71, 1980.
204. Buckett, W. R., Irreversible inhibitors of GABA transaminase induce antinociceptive effects and potentiate morphine, *Neuropharmacology*, 19, 715, 1980.
205. Groppe, G. and Kuschinsky, K., Stimulation and inhibition of serotoninergic mechanisms in rat brain: alterations of morphine effects of striatal dopamine metabolism and on motility, *Neuropharmacology*, 14, 659, 1975.
206. Duncan, C. and Spencer, P. S. J., An interaction between morphine and fenfluramine in mice, *J. Pharm. Pharmacol.*, 25, 124P, 1973.
207. Estler, C.-J., Effect of α- and β-adrenergic blocking agents and para-chlorophenylalanine on morphine- and caffeine-stimulated locomotor activity of mice, *Psychopharmacologia*, 28, 261, 1973.
208. Costall, B., Fortune, D. H., and Naylor, R. J., 5-HT antagonists inhibit neuroleptic and morphine antagonism of the hyperactivity induced by DA from the nucleus accumbens, *Br. J. Pharmacol.*, 60, 266, 1977.
209. Cicero, T. J., Meyer, E. R., and Smithloff, B. R., Alpha adrenergic blocking agents: antinociceptive activity and enhancement of morphine-induced analgesia, *J. Pharmacol. Exp. Ther.*, 189, 72, 1974.
210. Bhargava, N. and Way, E. L., Effect of 1-phenyl-3-(2-thiazolyl)-2-thiourea, a dopamine β-hydroxylase inhibitor, on morphine analgesia, tolerance, and physical dependence, *J. Pharmacol. Exp. Ther.*, 190, 165, 1974.
211. Watanabe, K., Matsui, Y., and Iwata, H., Enhancement of the analgesic effect of morphine by sodium diethyldithiocarbamate in rats, *Experientia*, 25, 950, 1969.
212. El Hawari, A. M., Cianflone, D., and Sharkawi, M., Effects of disulfiram on pharmacological activity toxicity and fate of morphine in rats, *Pharmacologist*, 19, 142, 1977.
213. Davis, W. M. and Smith, T. P., Morphine enhancement of shuttle avoidance prevented by α-methyltyrosine, *Psychopharmacologia*, 44, 95, 1975.
214. Wray, S. R., Possible catecholamine mediation of levallorphan-induced behavioral disruption in rats, *Psychopharmacologia*, 30, 251, 1973.
215. Blundell, C. and Slater, P., The effect of 6-hydroxydopamine on the antinociceptive action of analgesics in mice, *J. Pharm. Pharmacol.*, 29, 306, 1977.
216. Slater, P. and Blundell, C., The effects of a permanent and selective depletion of brain catecholamines on the antinociceptive action of morphine, *NaunynSchmiedeb. Arch. Pharmacol.*, 305, 227, 1978.
217. Ayhan, I. H., Effect of 6-hydroxydopamine on morphine analgesia, *Psychopharmacologia*, 25, 183, 1972.
218. Samanin, R. and Bernasconi, S., Effects of intraventriculary injected 6-OH dopamine or midbrain raphe lesion on morphine analgesia in rats, *Psychopharmacologia*, 25, 175, 1972.
219. Elchisak, M. A. and Rosecrans, J. A., Effect of central catecholamine depletions by 6-hydroxydopamine on morphine antinociception in rats, *Res. Commun. Chem. Pathol. Pharm.*, 6, 349, 1973.
220. York, J. L. and Maynert, E. W., Alteration in morphine analgesia produced by chronic deficits of brain catecholamines or serotonin: role of analgesimetric procedure, *Psychopharmacology*, 56, 119, 1978.
221. Courvoisier, S., Fournel, J., Ducrot, R., Kolsky, M., and Koetschet, P., Propriétés pharmacodynamiques de chlorhydrate de chloro-3 (diméthylamino-3′propyl)-10 phénothiazine (4.560 R.P.). Etude expérimentale d'un nouveau corps utilisé dans l'anesthésie potentialisée et dans l'hibernation artificielle, *Arch. Int. Pharmacodyn.*, 92, 305, 1953.
222. Kopera, J. and Armitage, A. K., Comparison of some pharmacological properties of chlorpromazine, promethazine, and pethidine, *Br. J. Pharmacol.*, 9, 392, 1954.
223. Dunai-Kovács, Z. and Székely, J. I., Effect of apomorphine on the antinociceptive activity of morphine, *Psychopharmacology*, 53, 65, 1977.

224. Spaulding, T. C., Fielding, S., Venfro, J. J., and Lal, H., Antinociceptive activity of clinidine and its potentiation of morphine analgesia, *Eur. J. Pharmacol.*, 58, 19, 1979.
225. Hesse, E., Zur biologischen Wertbestimmung der Analgetika und ihrer Kombinationen, *Arch. Exp. Pathol. Pharm.*, 158, 233, 1930.
226. Lal, H., Gianutsos, G., and Puri, S. K., A comparison of narcotic analgesics with neuroleptics on behavioral measures of dopaminergic activity, *Life Sci.*, 17, 29, 1975.
227. Lucot, J. B., McMilan, D. E., and Leander, J. D., The behavioral effects of chlorpromazine alone and in combination with different morphine treatments in rats, *J. Pharmacol. Exp. Ther.*, 208, 67, 1978.
228. Tarnawa, I. and Székely, J. I., Effect of morphine in combination with chlorpromazine and haloperidol on operant behavior, *Pharmacology*, 21, 186, 1980.
229. McKenzie, G. M. and Sadof, M., Effects of morphine and chlorpromazine on apomorphine-induced stereotyped behavior, *J. Pharm. Pharmacol.*, 26, 280, 1974.
230. Sturtevant, F. M. and Drill, V. A., Tranquilizing drugs and morphine mania in cats, *Nature (London)*, 179, 1253, 1957.
231. Winter, J. C., Proparanolol and morphine: a lethal interaction, *Arch. Int. Pharmacodyn.*, 212, 195, 1974.
232. Spiehler, V., Fairhurst, A. S., and Randall, L. O., The interaction of phenoxybenzamine with the mouse brain opiate receptor, *Mol. Pharmacol.*, 14, 587, 1978.
233. Cicero, T. J., Wilcox, C. E., and Meyer, E. R., Effect of α-adrenergic blockers on naloxone-binding in brain, *Biochem. Pharmacol.*, 23, 2349, 1974.
234. Elliott, H. W., Spiehler, V., and Navarro, G., Effect of naloxone on nociceptive activity of phenoxybenzamine, *Life Sci.*, 19, 1637, 1976.
235. Ireson, J. D., A comparison of the antinociceptive actions of cholinomimetic and morphine-like drugs, *Br. J. Pharmacol.*, 40, 92, 1970.
236. Saxena, P. N. and Gupta, G. P., Potentiating effects of prostigmine on morphine induced analgesia, *Indian J. Med. Res.*, 45, 319, 1957.
237. Parkes, M. W., Investigation of new chemical componens for spasmolytic activity, *Phys. Thesis, London Univ.*, 1957.
238. Howes, J. F., Harris, L. S., Dewey, W. L., and Voyda, C. A., Brain acetylcholine levels and inhibition of the tail flick reflex in mice, *J. Pharmacol. Exp. Ther.*, 169, 23 1969.
239. Kaakkola, S. and Ahtee, L., Effect of muscarinic cholinergic drugs on morphine-induced catalepsy, antinociception and changes in brain dopamine metabolism, *Psychopharmacology*, 52, 7, 1977.
240. Geller, E. B., Durlofsky, L., Harakal, C., Cowan, A., and Adler, M. W., Pentobarbital does not influence the antinociceptive effects of morphine in naive or morphine-tolerant rats, *Pharmacologist*, 19, 142, 1977.
241. Shannon, H. E., Holtzman, S. G., and Davis, D. G., Interaction between narcotic analgesics and benzodiazepine derivatives on behavior in the mouse, *J. Pharmacol. Exp. Ther.*, 199, 389, 1976.
242. Lee, R. J. and Spencer, P. S. J., Effect of tricyclic antidepressants on analgesic activity in laboratory animals, *Postgrad. Med. J.*, 56(Suppl. 1), 19, 1980.
243. Saarnivaara, L. and Mattila, M. J., Comparison of tricyclic antidepressants in rabbits: antinociception and potentiation of the noradrenaline pressor responses, *Psychopharmacologia*, 35, 221, 1974.
244. Székely, J. I. and Dunai-Kovács, Z., in preparation.
245. Breuker, E., Dingledine, R., and Iversen, L. L., Evidence for naloxone and opiates as GABA antagonists, *Br. J. Pharmacol.*, 58, 458, 1976.
246. Bhattacharya, S. K., Reddy, P. K. S. P., Debnath, P. K., and Sanyal, S. K., Potentiation of antinociceptive action of morphine by prostaglandin E_1 in albino rats, *Clin. Exp. Pharmacol. Physiol.*, 2, 353, 1975.
247. Ferri, S., Santagostino, A., Braga, P. C., and Galatulas, I., Decreased antinociceptive effect of morphine in rats treated intraventricularly with prostaglandin E_1, *Psychopharmacologia*, 39, 231, 1974.
248. Herz, A., Albus, K., Metyš, J., Schubert, P., and Teschemacher, H. J., On the central sites for the antinociceptive action of morphine and fentanyl, *Neuropharmacology*, 9, 539, 1970.
249. Pert, A. and Yaksh, T. L., Sites of morphine-induced analgesia in the primate brain: relation to pain pathways, *Brain Res.*, 80, 135, 1974.
250. Lotti, V. J., Lomax, P., and George, R., Temperature responses in the rat following intracerebral microinjection of morphine, *J. Pharmacol. Exp. Ther.*, 150, 135, 1965.
251. Foster, R. S., Jenden, D. J., and Lomax, P., A comparison of the pharmacological effects of morphine and N-methyl-morphine, *J. Pharmacol. Exp. Ther.*, 157, 185, 1967.
252. Sharpe, L. G., Garnett, J. E., and Cicero, T. J., Analgesia and hyperreactivity produced by intracranial microinjections of morphine into the periacqueductal gray matter of the rat, *Behav. Biol.*, 11, 303, 1974.

253. Bergmann, F., Chaimovitz, M., and Pasternak, V., Dual action of morphine and related drugs on compulsive gnawing of rats, *Psychopharmacologia*, 46, 87, 1976.
254. Jacquet, Y. F., Carol, M., and Russel, I. S., Morphine-induced rotation in naive, nonlesioned rats, *Science*, 192, 261, 1976.
255. Jacquet, Y. F., Klee, W. A., Rice, K. C., Iijima, I., and Minamikawa, J., Stereospecific and nonstereospecific effects of /+/- and /-/-morphine: evidence for a new class of receptors? *Science*, 198, 842, 1977.
256. Yaksh, T. L., Yeung, J. C., and Rudy, T. A., Systematic examination in the rat of brain sites sensitive to the direct application of morphine: observation of differential effects within the periaqueductal gray, *Brain Res.*, 114, 83, 1976.
257. Lewis, V. A. and Gebhart, G. F., Evaluation of the periaqueductal gray (PAG) as a morphine-specific locus of action and examination of morphine-induced and stimulation-produced analgesia at coincident PAG loci, *Brain Res.*, 124, 283, 1977.
258. Van Ree, J. M., Multiple brain sites involved in morphine antinociception, *J. Pharm. Pharmacol.*, 29, 765, 1977.
259. Dill, R. E. and Costa, E., Behavioral dissociation of the enkephalinergic systems of nucleus accumbens and nucleus caudatus, *Neuropharmacology*, 16, 323, 1977.
260. Chance, W. T. and Rosecrans, J. A., Inhibition of drinking by intrahypothalamic administration of morphine, *Nature (London)*, 270, 167, 1977.
261. Takagi, H., Satoh, M., Akaike, A., Shibata, T., and Kuraishi, Y., The nucleus reticularis gigantocellularis of medulla oblongata is a highly sensitive site in the production of morphine analgesia in the rat, *Eur. J. Pharmacol.*, 45, 91, 1977.
262. Costall, B., Fortune, D. H., and Naylor, R. J., Involvement of mesolimbic and extrapyramidal nuclei in the motor depressant action of narcotic drugs, *J. Pharm. Pharmacol.*, 30, 566, 1978.
263. Costall, B., Fortune, D. H., and Naylor, R. J., The induction of catalepsy and hyperactivity by morphine administered directly into the nucleus accumbens of rats, *Eur. J. Pharmacol.*, 49, 49, 1978.
264. Moroni, F., Cheney, D. L., and Costa, E., The turnover rate of acetylcholine in brain nuclei of rats injected intraventricularly and intraseptally with alpha- and beta-endorphin, *Neuropharmacology*, 17, 191, 1978.
265. Dickenson, A. H., Oliveras, J. L., and Besson, J. M., Role du noyau Raphe Magnus dans l'analgesie morphinique: Etudes par microinjections intracerebrales chez le Rat, *C. R. Acad. Sci. Ser. D.*, 287, 955, 1978.
266. Dickenson, A. H., Oliveras, J. L., and Besson, J. M., Role of the nucleus raphe magnus in opiate analgesia as studied by the microinjection technique in the rat, *Brain Res.*, 170, 95, 1979.
267. Akaike, A., Shibata, T., Satoh, M., and Takagi, H., Analgesia induced by microinjection of morphine into, and electrical stimulation of the nucleus reticularis paragigantocellularis of rat medullar oblongata, *Neuropharmacology*, 17, 775, 1978.
268. Jurna, I. and Heinz, G., Anti-nociceptive effect of morphine, opioid analgesics and haloperidol injected into the caudate nucleus of the rat, *Naunyn-Schmiedeb. Arch. Pharmacol.*, 309, 145, 1979.
269. Jacquet, Y. F., Dual mechanisms mediating opiate effects? *Science*, 205, 425, 1979.
270. Metys, Von J., Metysova, J., Wagner, N., Schondorf, N., und Herz, A., Hemmung nociceptiver Reaktionen durch Cholinomimetika und durch Morphin nach intraventrikularer und intracerebraler Injektion, *Arznm. - Forsch. (Drug Res.,)*, 19, 432, 1969.
271. Yaksh, T. L. and Tyce, G. M., Microinjection of morphine into the periaqueductal gray evokes the release of serotonin from spinal cord, *Brain Res.*, 171, 176, 1979.
272. Yaksh, T. L., Direct evidence that spinal serotonin and noradrenaline terminals mediate the spinal antinociceptive effects of morphine in the periaqueductal gray, *Brain Res.*, 160, 180, 1979.
273. Yaksh, T. L., Wilson, P. R., Kaiko, R. F., and Inturrisi, C. E., Analgesia produced by a spinal action of morphine and effects upon parturition in the rat, *Anesthesiology*, 51, 386, 1979.
274. Jones, R. D. M. and Jones, J. G., Intrathecal morphine:naloxone reverses respiratory depression but not analgesia, *Br. Med. J.*, 281, 645, 1980.
275. Reynolds, D. V., Surgery in the rat during electrical analgesia induced by focal brain stimulation, *Science*, 164, 444, 1969.
276. Willer, C. J. and Bussel, B., Possible explanation for analgesia mediated by direct spinal effect of morphine, *Lancet*, 1, 158, 1980.
277. Costall, B., Fortune, D. H., and Naylor, R. J., Biphasic changes in motor behavior following morphine injection into the nucleus accumbens, *Br. J. Pharmacol.*, 57, 423, 1976.
278. Costall, B. and Naylor, R. J., A role for the amygdala in the development of the cataleptic and stereotypic actions of the narcotic agonists and the antagonists in the rat, *Psychopharmacologia*, 35, 203, 1974.
279. Liljequist, R. and Mattila, M. J., Codeine-induced facilitation of memory functions, *Br. J. Clin. Pharmacol.*, 4, 654, 1977.

280. Mondadori, C. and Waser, P. G., Improvement of retention by posttrial morphine, *Experientia*, 34, 902, 1978.
281. White, N., Major, R., and Siegel, J., Effects of morphine on one-trial appetitive learning, *Life Sci.*, 23, 19, 1967.
282. Gallagher, M. and Kapp, B. S., Manipulation of opiate activity in the amygdala alters memory processes, *Life Sci.*, 23, 1973, 1978.
283. Geller, I. and Seifter, J., The effects of meprobamate, barbiturates, d-amphetamine and promazine on experimentally induced conflict in the rat, *Psychopharmacologia*, 1, 482, 1960.
284. Hill, H. E., Belleville, R. E., Pescor, F. T., and Winkler, A., Comparative effects of methadone, meperidine and morphine on conditioned supression, *Arch. Int. Pharmacodyn.*, 163, 341, 1966.
285. Gellert, V. F. and Sparber, S. B., Effects of morphine withdrawal on food competition hierarchies and fighting behavior in rats, *Psychopharmacology*, 60, 165, 1979.
286. Thompson, T., Trombley, J., Luke, D., and Lott, D., Effects of morphine on behavior maintained by four simple food-reinforcement schedules, *Psychopharmacologia*, 17, 182, 1970.
287. McMillan, D. E., Wolf, P. S., and Carchman, R. A., Antagonism of the behavioral effect of morphine and methadone by narcotic antagonists in the pigeon, *J. Pharmacol. Exp. Ther.*, 175, 443, 1970.
288. Banerjee, U., Acquisition of conditioned avoidance response in rats under the influence of addicting drugs, *Psychopharmacologia*, 22, 133, 1971.
289. Herman, S. J., Freeman, B. J., and Ray, O. S., The effects of multiple injections of morphine sulfate on shuttle box behavior in the rat, *Psychopharmacologia*, 26, 146, 1972.
290. Davis, W. M. and Smith, T. P., Morphine enhancement of shuttle avoidance prevented by α-methyltyrosine, *Psychopharmacologia*, 44, 95, 1975.
291. Cook, L. and Widley, E., Behavioral effects of some psychopharmacological agents, *Ann. N.Y. Acad. Sci.*, 66, 740, 1957.
292. Verhave, T., Owen, J. E., Jr., and Robbins, E. B., The effect of morphine sulfate on avoidance and escape behavior, *J. Pharmacol. Exp. Ther.*, 125, 248, 1959.
293. Holtzman, S. G., Tolerance to the stimulant effects of morphine and pentazocine on avoidance responding in the rat, *Psychopharmacologia*, 39, 23, 1974.
294. Davis, W. M., Holbrook, J. M., and Babbini, M., Differential effects of morphine on active avoidance as a function of pre-drug performance, *Pharm. Res. Comm.*, 5, 47, 1973.
295. Satinder, K. P., Differential effects of morphine on two-way avoidance in selectively bred rat strains, *Psychopharmacology*, 48, 235, 1976.
296. Ageel, A. M., Chin, L., Trafton, C. L., Jones, B. C., and Picchioni, A. L., Acute effects of morphine and chlorpromazine on the acquisition of shuttle box conditioned avoidance response, *Psychopharmacologia*, 46, 311, 1976.
297. Giarman, N. J. and Pepeu, G., Drug-induced changes in brain acetylcholine, *Br. J. Pharmacol.*, 19, 226, 1962.
298. Michell, J. F., The spontaneous and evoked release of acetylcholine from the cerebral cortex, *J. Physiol. (London)*, 165, 98, 1963.
299. Hano, K., Kaneto, H., Kakunaga, T., and Moribayashi, N., Pharmacological studies of analgesics. VI. The administration of morphine and changes in acetylcholine metabolism mouse brain, *Biochem. Pharmacol.*, 13, 441, 1964.
300. Beleslin, D. and Polak, R. L., Depression by morphine and chloralose of acetylcholine release from the cat's brain, *J. Physiol. (London)*, 177, 411, 1965.
301. Crossland, J. and Slater, P., The effect of some drugs on the "free" and "bound" acetylcholine content of rat brain, *Br. J. Pharmacol.*, 33, 42, 1968.
302. Beani, L., Bianchi, C., Santinoceto, L., and Marchetti, P., The cerebral acetylcholine release in conscious rabbits with semi-permanently implanted epidural cups, *Int. J. Neuropharmacol.*, 7, 469, 1968.
303. Richter, J. A. and Goldstein, A., Effects of morphine and levorphanol on brain acetylcholine content in mice, *J. Pharmacol. Exp. Ther.*, 175, 685, 1970.
304. Jhamandas, K., Pinsky, C., and Phillis, J. W., Effects of morphine and its antagonists on release of cerebral cortical acetylcholine, *Nature (London)*, 228, 176, 1970.
305. Jhamandas, K., Phillis, J. W., and Pinsky, C., Effects of narcotic analgesics and antagonists on the *in vivo* release of acetylcholine from the cerebral cortex of the cat, *Br. J. Pharmacol.*, 43, 53, 1971.
306. Sharkawi, M., Effects of morphine and pentobarbitone on acetylcholine synthesis by rat cerebral cortex, *Br. J. Pharmacol.*, 40, 86, 1970.
307. Datta, K., Thal, L., and Wajda, I. J., Effects of morphine on choline acetyltransferase levels in the caudate nucleus of the rat, *Br. J. Pharmacol.*, 41, 84, 1971.
308. Domino, E. F., and Wilson, A. E., Psychotropic drug influences on brain acetylcholine utilization, *Psychopharmacologia*, 25, 291, 1972.

309. Domino, E. F. and Wilson, A., Effects of narcotic analgesic agonists and antagonists on rat brain acetylcholine, *J. Pharmacol. Exp. Ther.*, 184, 18, 1973.
310. Matthews, J. D., Labrecque, G., and Domino, E. F., Effects of morphine, nalorphine and naloxone on neocortical release of acetylcholine in the rat, *Psychopharmacologia*, 29, 113, 1973.
311. Cheney, D. L., Trabucchi, M., Racagni, G., Wang, C., and Costa, E., Effects of acute and chronic morphine on regional rat brain acetylcholine turnover rate, *Life Sci.*, 15, 1977, 1974.
312. Mullin, W. J. and Phillis, J. W., Acetylcholine release from the brain of unanaesthetized cats following habituation to morphine and during precipitation of the abstinence syndrome, *Psychopharmacologia*, 36, 85, 1974.
313. Labrecque, G. and Domino, E. F., Tolerance to and physical dependence on morphine: relation to neocortical acetylcholine release in the cat, *J. Pharmacol. Exp. Ther.*, 191, 189, 1974.
314. Mehta, V. L., Cholinergic mechanisms in narcotic analgesics, *Neuropharmacology*, 14, 893, 1975.
315. Torda, C. and Wolff, H. G., Effect of alkaloids on acetylcholine synthesis, *Arch. Biochem.*, 10, 247, 1947.
316. Morris, R. W., Effects of drugs on the biosynthesis of acetylcholine: pentobarbital, morphine and morphinan derivatives, *Arch. Int. Pharmacodyn.*, 133, 326, 1961.
317. Domino, E. F., Vasko, M. R., and Wilson, A. E., Mixed depressant and stimulant actions of morphine and their relationhip to brain acetylcholine, *Life Sci.*, 18, 361, 1976.
318. Zsilla, G., Cheney, D. L., Racagni, G., and Costa, E., Correlation between analgesia and the decrease of acetylcholine turnover rate in cortex and hippocampus elicited by morphine, meperidine, viminol R_2 and azidomorphine, *J. Pharmacol. Exp. Ther.*, 199, 662, 1976.
319. Zsilla, G., Racagni, G., Cheney, D. L., and Costa, E., Constant rate infusion of deuterated phosphorylcholine to measure the effects of morphine on acetylcholine turnover rate in specific nuclei of rat brain, *Neuropharmacology*, 16, 25, 1977.
320. Moroni, F., Cheney, D. L., and Costa, E., Inhibition of acetylcholine turnover in rat hippocampus by intraseptal injections of β-endorphin and morphine, *Naunyn-Schmiedeb. Arch. Pharmacol.*, 299, 149, 1977.
321. Goldstein, T. R., Effect of acute and chronic administration of morphine on cholinergic mechanisms in the central nervous system, *Dissertation Abstr. Intern.*, 39, 2244, 1978.
322. Vizi, E. S., Hársing, L. G., Jr., and Knoll, J., Presynaptic inhibition leading to disinhibition of acetylcholine release from interneurones of the caudate nucleus: effects of dopamine β-endorphin and D-Ala²-Pro⁵-enkephalinamide, *Neuroscience*, 2, 953, 1977.
323. Sharkawi, M. and Schulman, M. P., Inhibition by morphine of the release of (^{14}C) acetylcholine from rat brain cortex slices, *J. Pharm. Pharmacol.*, 21, 546, 1969.
324. Yaksh, T. L. and Yamamura, H. I., Depression by morphine of the resting and evoked release of (^{3}H)-acetylcholine from the cat caudate nucleus in vivo, *Neuropharmacology*, 16, 227, 1977.
325. Rujirekagulwat, T., Redburn, D. A., Cope, T., and Rosefeld, G. C., Morphine-induced stimulation of high affinity choline uptake in rat brain striatum, *Pharmacologist*, 18, 213, 1976.
326. Way, E. L., Loh, H. H., and Shen, F.-H., Morphine tolerance, physical dependence, and synthesis of brain 5-hydroxytryptamine, *Science*, 162, 1290, 1968.
327. Sanfacon, G. and Labrecque, G., Acetylcholine antirelease effect of morphine and its modification by calcium, *Psychopharmacology*, 55, 151, 1977.
328. Kakunaga, T., Kaneto, H., and Hano, K., Pharmacologic studies on analgesics-VII. Significance of the calcium ion in morphine analgesia, *J. Pharmacol. Exp. Ther.*, 153, 134, 1966.
329. Kaneto, H., Inorganic ions: the role of calcium, in *Narcotic Drugs: Biochemical Pharmacology*, Clouet, D. H., Ed., Plenum Press, New York, 1971, 300.
330. Harris, R. A., Loh, H. H., and Way, E. L., Antinociceptive effects of lanthanum and cerium in nontolerant and morphine tolerant-dependent animals, *J. Pharmacol. Exp. Ther.*, 196, 288, 1976.
331. Schmidt, W. K. and Way, E. L., Hyperalgesic effects of divalent cations and antinociceptive effects of a calcium chelator in naive and morphine-dependent mice, *J. Pharmacol. Exp. Ther.*, 212, 22, 1980.
332. Jhamandas, K., Sawynok, J., and Sutak, M., Antagonism of morphine action on brain acetylcholine release by methylxanthines and calcium, *Eur. J. Pharmacol.*, 49, 309, 1978.
333. Burks, T. F. and Long, J. P., Release of intestinal 5-hydroxytryptamine by morphine and related agents, *J. Pharmacol. Exp. Ther.*, 156, 267, 1967.
334. Snow, A. and Dewey, W. L., The effects of morphine and opioid peptides on whole mouse brain serotonergic systems, *Pharmacologist*, 20, 268, 1978.
335. Yarbrough, G. G., Buxbaum, D. M., and Sanders-Bush, E., Increased serotonin turnover in the acutely morphine-treated rat, *Life Sci.*, 10, 977, 1971.
336. Yarbrough, G. G., Buxbaum, D. M., and Sanders-Bush, E., Increased serotonin turnover in acutely morphine-treated mice, *Biochem. Pharmacol.*, 21, 2667, 1972.

337. Yarbrough, G. G., Buxbaum, M., and Sanders-Bush, E., Biogenic amines and narcotic effects. II. Serotonin turnover in the rat after acute and chronic morphine administration, *J. Pharmacol. Exp. Ther.*, 185, 328, 1973.
338. Haubrich, D. R. and Blake, D. E., Modification of serotonin metabolism in rat brain after acute or chronic administration of morphine, *Biochem. Pharmacol.*, 22, 2753, 1973.
339. Goodlet, I. and Sugrue, M. F., Effect of acutely administered analgesic drugs on rat brain serotonin turnover, *Eur. J. Pharmacol.*, 29, 241, 1974.
340. Pérez-Cruet, J., Thoa, N. B., and Ng, L. K. Y., Acute effects of heroin and morphine on newly synthesized serotonin in rat brain, *Life Sci.*, 17, 349, 1975.
341. Sawa, A. and Oka, T., Effects of narcotic analgesics on serotonin metabolism in brain of rats and mice, *Jpn. J. Pharmacol.*, 26, 599, 1976.
342. Larson, A. A. and Takemori, A. E., Effects of narcotics on tbe uptake of serotonin precursors by the rat brain, *J. Pharmacol. Exp. Ther.*, 200, 216, 1977.
343. Larson, A. A., The effect of narcotic analgesics on the transport and fate of serotonin precursors in the brain of rats, *Dissertation Abstr. Intern. B*, 38, 1160, 1977.
344. Warwick, R. O., Bousquet, W. F., and Craig-Schnell, R., Effect of acute and chronic morphine treatment on serotonin uptake into rat hypothalamic synaptosomes, *Pharmacology*, 15, 415, 1977.
345. Messing, R. B., Flinchbaugh, C., and Waymire, J. C., Changes in brain tryptophan and tyrosine following acute and chronic morphine administration, *Neuropharmacology*, 17, 391, 1978.
346. Shiomi, H., Murakami, H., and Takagi, H., Morphine analgesia and the bulbospinal serotoninergic system: increase in concentration of 5-hydroxyindolacetic acid in the rat spinal cord with analgesics, *Eur. J. Pharmacol.*, 52, 335, 1978.
347. Weil-Fugazza, J., Godefroy, F., and Besson, J. M., Effect of morphine on the metabolism of 5-HT at spinal and supraspinal levels in normal and arthritic rats, *Proc. 7th Int. Congr. Pharm.*, Paris, France, 1978, Abstract, 119.
348. Aiello-Malmberg, P., Bartolini, A., Bartolini, R., and Galli, A., Effects of morphine, physostigmine and raphe nuclei stimulation on 5-hydroxytryptamine release from the cerebral cortex of the cat, *Br. J. Pharmacol.*, 65, 547, 1979.
349. Wang, J. K., Antinociceptive effect of intrathecally administered serotonin, *Anesthesiology*, 47, 269, 1977.
350. Yaksh, T. L. and Wilson, P. R., Spinal seotonin terminal system mediates antinociception, *J. Pharmacol. Exp. Ther.*, 208, 446, 1979.
351. Deakin, J. F. W. and Dostrovsky, J. O., Involvement of the periaqueductal grey matter and spinal 5-hydroxytryptaminergic pathways in morphine analgesia: effects of lesions and 5-hydroxytryptamine depletion, *Br. J. Pharmacol.*, 63, 159, 1978.
352. Vogt, M., The concentration of sympathin in different parts of the central nervous system under normal conditions and after the administration of drugs, *J. Physiol. (London)*, 123, 451, 1954.
353. Maynert, E. W. and Klingman, G. I., Tolerance to morphine. I. Effects on catecholamines in the brain and adrenal glands, *J. Pharmacol. Exp. Ther.*, 135, 285, 1962.
354. Gunne, L.-M., Fuxe, K., and Jonsson, J., Effects of morphine intoxication on brain catecholamine neurons, *Eur. J. Pharmacol.*, 5, 338, 1969.
355. Clouet, D. H. and Ratner, M., Catecholamine biosynthesis in brains of rats treated with morphine, *Science*, 168, 854, 1970.
356. Smith, C. B., Sheldon, M. I., and Bednarczyk, J. H., Willarreal, J. E., Morphine-induced increases in the incorporation of ^{14}C-tyrosine into ^{14}C-dopamine and ^{14}C-norepinephrine in the mouse brain: antagonism by naloxone and tolerance, *J. Pharmacol. Exp. Ther.*, 180, 547, 1972.
357. Fukui, K. and Takagi, H., Effect of morphine on the cerebral contents of metabolites of dopamine in normal and tolerant mice: its possible relation to analgesic action, *Br. J. Pharmacol.*, 44, 45, 1972.
358. Fukui, K., Shiomi, H., and Takagi, H., Effect of morphine on tyrosine hydroxylase activity in mouse brain, *Eur. J. Pharmacol.*, 19, 123, 1972.
359. Gauchy, C., Agid, Y., Glowinski, J., and Cheramy, A., Acute effects of morphine on dopamine synthesis and release and tyrosine metabolism in the rat striatum, *Eur. J. Pharmacol.*, 22, 311, 1973.
360. Marshall, I. and Smith, C. B., Acute and chronic morphine treatment and the hydroxylation of (1-^{14}C)-1-tyrosine in the mouse brain, *Br. J. Pharmacol.*, 50, 428, 1974.
361. Sugrue, M. F., The effects of acutely administered analgesics on the turnover of noradrenaline and dopamine in various regions of the rat brain, *Br. J. Pharmacol.*, 52, 159, 1974.
362. Freye, E. and Kuschinsky, K., Effects of fentanyl and droperidol on the dopamine metabolism of the rat striatum, *Pharmacology*, 14, 1, 1976.
363. Moleman, P. and Bruinvels, J., Differential effect of morphine on dopaminergic neurons in frontal cortex and striatum of the rat, *Life Sci.*, 19, 1277, 1976.
364. Bloom, A. S., Dewey, W. L., Harris, L. S., and Brosius, K. K., The correlation between antinociceptive activity of narcotics and their antagonists as measured in the mouse tail flick test and increased synthesis of brain catecholamines, *J. Pharmacol. Exp. Ther.*, 198, 33, 1976.

365. Westerink, B. H. C. and Korf, J., Regional rat brain levels of 3,4-dihydroxyphenylacetic acid and homovanillic acid: concurrent fluorometric measurement and influence of drugs, *Eur. J. Pharmacol.*, 38, 281, 1976.
366. Kuraishi, Y., Fukui, K., Shiomi, H., Akaike, A., and Takagi, H., Microinjection of opioids into the nucleus reticularis gigantocellularis of the rat: analgesia and increase in the normetanephrine level in the spinal cord, *Biochem. Pharmacol.*, 27, 2756, 1978.
367. Kameyama, T., Nabeshima, T., Ukai, M., and Yamaguchi, K., Morphine-induced Straub tail reaction and spinal catecholamine metabolite content: Antagonism of naloxone to morphine-induced effects in mice, *Chem. Pharm. Bull.*, 26, 2615, 1978.
368. Bensemana, D. and Gascon, A. L., Relationship between analgesia and turnover of brain biogenic amines, *Can. J. Physiol. Pharmacol.*, 56, 721, 1978.
369. Deyo, S. N., Swift, R. M., and Miller, R. J., Morphine and endorphins modulate dopamine turnover in rat median eminence, *Proc. Natl. Acad. Sci. USA*, 76, 3006, 1979.
370. Moleman, P. and Bruinvels, J., Morphine-induced striatal dopamine efflux depends on the activity of nigrostriatal dopamine neurones, *Nature (London)*, 281, 686, 1979.
371. Sharpe, L. G., Garnett, J. E., and Cicero, T. J., Analgesia and hyperactivity produced by intracranial microinjections of morphine into the periaqueductal gray matter of the rat, *Behav. Biol.*, 11, 303, 1974.
372. Hitzeman, R. J. and Loh, H. H., Effect of morphine on the transport of dopamine into mouse brain slices, *Eur. J. Pharmacol.*, 21, 121, 1973.
373. Carmichael, F. J. and Israel, Y., In vitro inhibitory effects of narcotic analgesics and other psychotropic drugs on the uptake of norepinephrine in mouse brain tissue, *J. Pharmacol. Exp. Ther.*, 186, 253, 1973.
374. Celsen, B. and Kuschinsky, K., Effects of morphine on kinetics of ^{14}C-dopamine in rat striatal slices, *Naunyn-Schmiedeb. Arch. Pharmacol.*, 284, 159, 1974.
375. Bosse, A., In vitro studies in synaptosomes of rat striatum about the effects of morphine on uptake and release of C-14-dopamine, *Naunyn-Schmiedeb. Arch. Pharmacol.*, 297, 52R, 1977.
376. Arbilla, S. and Langer, S. Z., Morphine and β-endorphin inhibit release of noradrenaline from cerebral cortex but not of dopamine from rat striatum, *Nature (London)*, 271, 559, 1978.
377. Eibergen, R. D. and Carlson, K. R., Behavioral evidence for dopaminergic supersensitivity following chronic treatment with methadone or chlorpromazine in the guinea pig, *Psychopharmacology*, 48, 139, 1976.
378. Rae, G. A., Neto, J. P., and Moraes, S. D., Noradrenaline supersensitivity of the mouse vas deferens after long-term treatment with morphine, *J. Pharm. Pharmacol.*, 29, 310, 1977.
379. Andén, N. E. and Andén - Grabowska, M., Morphine-induced changes in striatal dopamine mechanisms not evoked from the dopamine nerve terminals, *J. Pharm. Pharmacol.*, 30, 732, 1978.
380. Székely, J. I., Horváth, K., Markovits, J., and Miglécz. E., in preparation.
381. Sanghvi, I. S. and Gershon, S., Brain calcium and morphine action, *Biochem. Pharmacol.*, 26, 1183, 1977.
382. Mulé, S. J., Morphine and the incorporation of P^{32} into brain phospholipids of nontolerant, tolerant and abstinent guinea pigs, *J. Pharmacol. Exp. Ther.*, 156, 92, 1967.
383. Feldberg, W. and Shaligram, S. V., The hyperglycaemic effect of morphine, *Br. J. Pharmacol.*, 46, 602, 1972.
384. Ross, D. H., Medina, M. A., and Cardenas, L. H., Morphine and ethanol: selective depletion of regional brain calcium, *Science*, 186, 63, 1974.
385. Bonnet, K. A., Regional alterations in cyclic nucleotide levels with acute and chronic morphine treatment, *Life Sci.*, 16, 1877, 1975.
386. Iwatsubo, K. and Clouet, D. H., Dopamine-sensitive adenylate cyclase of the caudate nucleus of rats treated with morphine or haloperidol, *Biochem. Pharmacol.*, 24, 1499, 1975.
387. Harris, R. A., Harris, L. S., and Dunn, A., Effect of narcotic drugs on ribonucleic acid and nucleotide metabolism in mouse brain, *J. Pharmacol. Exp. Ther.*, 192, 280, 1975.
388. Cardenas, L. H. and Ross, D. H., Morphine induced calcium depletion in discrete regions of rat brain, *J. Neurochem.*, 24, 487, 1975.
389. Biggio, G. and Guidotti, A., Action of morphine in the regulation of cyclic GMP (cGMP) content of rat cerebellum, *Pharmacologist*, 18, 212, 1976.
390. Puri, S. K., Cochin, J., and Volicer, L., Effect of morphine sulfate on adenylate cyclase and phosphodiesterase activities in rat corpus striatum, *Life Sci.*, 16, 759, 1976.
391. Nakagawa, K. and Kuriyama, K., Morphine-induced changes of cyclic AMP metabolism and protein kinase activity in brain, *Jpn. J. Pharmacol.*, 26, 110, 1976.
392. Mazurkiewicz Kwilecki, I. and Henwood, R. W., Alterations in brain endogenous histamine levels in rats after chronic morphine treatment and morphine withdrawal, *Agents Actions*, 6, 402, 1976.

393. Cardenas, H. L. and Ross, D. H., Calcium depletion of synaptosomes after morphine treatment, *Br. J. Pharmacol.*, 57, 521, 1976.
394. Biggio, G., Guidotti, A., and Costa, E., On the mechanism of the decrease in cerebellar cyclic GMP content elicited by opiate receptor agonists, *Naunyn-Schmiedeb. Arch. Pharmacol.*, 296, 117, 1977.
395. Ford, D. H., Rhines, R. K., and Levi, M. A., Strain differences in the response to morphine on incorporation of ^{3}H-lysine into rat brain protein, *Acta Neurol. Scand.*, 55, 493, 1977.
396. Misra, A. I, Vadlamani, N. L., and Pontani, R. B., Differential effects of opiates on the incorporations of (^{14}C)-thiamine in the central nervous system of the rat, *Experientia*, 33, 372, 1977.
397. Harris, R. A., Yamamoto, H., Loh, H. H., and Way, E. L., Discrete changes in brain calcium with morphine analgesia, tolerance-dependence, and abstinence, *Life Sci.*, 20, 501, 1977.
398. Muraki, T., Tokunaga, Y., and Kato, B., Effects of morphine on the plasma cyclic AMP levels in male mice, *Proc. 7th Int. Congr. Pharm.*, Paris, France, 1978, Abstract, 358.
399. Muraki, T., Nakadate, T., Tokunaga, Y., and Kato, R., Effect of narcotic analgesics on plasma cyclic AMP levels in male mice, *Neuropharmacology*, 18, 623, 1979.
400. Salles, K. S. S., Colasanti, B. K., Craig, C. R., and Thomas, J. A., Involvement of brain cyclic AMP in the acute and chronic effects of morphine in the rat, *Pharmacology*, 17, 128, 1978.
401. Hui, F. W., Krikun, E., and Smith, A. A., Inhibition by d1-methadone of RNA and protein synthesis in neonatal mice: antagonism by naloxone or naltrexone, *Eur. J. Pharmacol.*, 49, 87, 1978.
402. Sethy, V. H. and Bombardt, P. A., Is GABA involved in analgesia, *Res. Commun. Chem. Pathol. Pharm.*, 19, 365, 1978.
403. Yamamoto, H., Harris, R. A., Loh, H. H., and Way, E. L., Effects of acute and chronic morphine treatments on calcium localization and binding in brain, *J. Pharmacol. Exp. Ther.*, 205, 255, 1978.
404. Santagostino, A., Giagnoni, G., Reina, R., Spadaro, C., and Ferri, S., Changes in liver tyrosine aminotransferase after acute and chronic administration of morphine in the rat, *Arch. Toxicol.*, 1, 331, 1978.
405. Kwong, T. H., Wong, S. C., and Yeung, D., Effect of morphine on biosynthetic pathways in rat liver, *IRCS Pharmacol.*, 8, 134, 1980.
406. Ferri, S., Reina, R. A., Spadaro, C., and Scoto, G. M., Morphine-induced changes in PG biosynthesis of rat placenta, *Proc. 7th Int. Congr. Pharm.*, Paris, France, 1978, Abstract, 750.
407. Guerrero-Munoz, F., Guerrero, M. L., Way, E. L., and Leake, C. D., Effect of acute and chronic morphine administration on synaptosomal Ca^{++} efflux, *Fed. Proc. Fed. Am. Soc. Exp. Biol.*, 37, 764, 1978.
408. Guerrero-Munoz, F., Cerreta, K. V., Guerrero, M. L., and Way, E. L., Effect of morphine on synaptosomal Ca^{++} uptake, *J. Pharmacol. Exp. Ther.*, 209, 132, 1979.
409. Racagni, G., Bruno, F., Juliano, E., and Paoletti, R., Differential sensitivity to morphine-induced analgesia and motor activity in two inbred strains of mice: behavioral and biochemical correlations, *J. Pharmacol. Exp. Ther.*, 209, 111, 1979.
410. Desiah, D. and Ho, I. K., Effect of morphine on mouse brain ATPase activities, *Biochem. Pharmacol.*, 26, 89, 1977.
411. Desiah, D. and Ho, I. K., Effects of acute and continuous morphine administration on catecholamine-sensitive adenosine triphosphatase in mouse brain, *J. Pharmacol. Exp. Ther.*, 208, 80, 1979.
412. Natsuki, R., Hitzemann, R. J., and Loh, H. H., Influence of morphine, β-endorphin and naloxone on the synthesis of phosphoinositides in the rat midbrain, *Res. Commun. Chem. Pathol. Pharm.*, 24, 233, 1979.
413. Lewis, S. J. and Fennessy, M. R., Effect of acute and chronic morphine on rat brain histamine levels, *Clin. Exp. Pharm. Physiol.*, 6, 223, 1979.
414. Lamb, R. G. and Dewey, W. L., Effect of morphine exposure on mouse liver triglyceride biosynthesis, *Pharmacologist*, 21, 226, 1979.
415. Chang, Y.-Y. H. and Ho, I. K., Effects of acute and continuous morphine administration on serum glutamate oxalacetate transaminase and glutamate pyruvate transaminase activities in the mouse, *Biochem. Pharmacol.*, 28, 1373, 1979.
416. Bogdanski, D. F., Tissari, A., and Brodie, B. B., Role of sodium, potassium, ouabain and reserpine in uptake, storage and metabolism of biogenic amines in synaptosomes, *Life Sci.*, 7, 419, 1968.
417. Yamamoto, H., Harris, R. A., Loh, H. H., and Way, E. L., Effects of morphine tolerance and dependence on Mg^{++} dependent ATPase activity of synaptic vesicles, *Life Sci.*, 20, 1533, 1977.
418. Mulé, S. J., Inhibition of phospholipid-facilitated calcium transport by central nervous system-acting drugs, *Biochem. Pharmacol.*, 18, 339, 1969.
419. Greenberg, S., Diecke, F. P. J., and Long, J. P., *In vitro* model for studying effects of morphine and nalorphine on ^{45}Ca-ganglioside binding, *J. Pharm. Sci.*, 61, 1471, 1972.
420. Clissiounis, N., Effect of adrenergic drugs on morphine-induced hyperglycemia, *Life Sci.*, 25, 391, 1979.
421. Gero, A., The action of opioid drugs on human plasma cholinesterase, *Life Sci.*, 25, 201, 1979.

INDEX

A

Aceperon, 47, 71
Acetylcholine
  behavioral effects studies, 47
  drug interaction studies, 63—64
  isolated organ studies, 8, 10—12
  naloxone antagonism studies, 30
  neurotransmitter turnover studies, 75—80, 87
N-Acetyl (Tyr¹)-β-EP, 6
ACTH, 6
ACTH/β-LPH precursor, 17—18
ACTH receptor, 48, 69
Adenosine triphosphatase, 90—92
Adenosine triphosphate, 88
Adenylate cyclase, 88—91
Adrenaline, 57—58, 61, 63, 84—85
Adrenergic receptor, 58, 63
α-Adrenergic receptor, 13, 47, 94
β-Adrenergic receptor, 63
α-Adrenergic receptor blockers, 48, 63, 71
β-Adrenergic receptor blockers, 48
Adrenergic receptor stimulants, 72
α-Adrenergic receptor stimulants, 41
Agonist, narcotic
  behavioral effects studies, 51
  mixed agonist-antagonists, see Mixed agonist-antagonists
  neurotransmitter turnover studies, 93
  opiate receptor affinity for, changes in, 37—38
  stimulus properties studies, 39—40
Agonist, opiate
  isolated organ studies, 10, 12—15
  mixed agonist-antagonists, see Mixed agonist-antagonists
  neurotransmitter turnover studies, 75, 84, 87
  receptor binding assays, 2
  stimulus properties studies, 40
D-Ala², Leu⁵-enkephalin, 14
D-Ala², D-Leu⁵-enkephalin, 14
Alcohol, 38
N-Allylnorphenazocine, 36
Amines, 53, 57, 60, 91
Amphetamine
  analgesic effects studies, 42
  behavioral effects studies, 51—52
  conditioned behavior studies, 74
  drug interaction studies, 61—62
  neurotransmitter turnover studies, 87—88
  stimulus properties studies, 38—40
AMT, see α-Methyl-p-tyrosine
Amygdala
  conditioned behavior studies, 73—74
  neurotransmitter turnover studies, 78
Analgesic effects
  behavioral effects studies, 45—47, 50—53
  drug interaction studies, 53—66
  in vivo studies, 42—45
  localized intracerebral application studies, 66—73
  naloxone antagonism studies, 31—37
  neurotransmitter turnover studies, 77—80, 82—83, 86—88, 92—93
  stimulus properties, narcotic analgesics, 39
Antagonism, naloxone, see Naloxone, antagonism
Antagonist, dopamine, 71
Antagonist, narcotic
  behavioral effects studies, 46—47
  mixed agonist-antagonists, see Mixed agonist-antagonists
  opiate receptor affinity for, changes in, 37—38
  pA₂ values against morphine, 33—38
  stimulus properties studies, 40
Antagonist, opiate
  behavioral effects studies, 51
  conditioned behavior studies, 74
  isolated organ studies, 10, 12—16
  mixed agonist-antagonists, see Mixed agonist-antagonists
  neurotransmitter turnover studies, 78, 84
  receptor binding assays, 2
Anticholinergic agents, 41
Antidepressants, 65
Antiinflammatory agents, 41—42
Antimuscarinic drugs, 64—65
Anxiolytic drugs, 74
Apomorphine
  behavioral effects studies, 51—52
  drug interaction studies, 61—63
  localized intracerebral application studies, 71
  neurotransmitter turnover studies, 81, 87—88
  stimulus properties studies, 40
Aqueduct, localized intracerebral application studies, 66—67
Arecoline, 64
Areflexia, 46
Ascending serotoninergic pathways, neurotransmitter turnover studies, 81—82
Aspirin, 41—42
ATP, see Adenosine triphosphate
ATPase, see Adenosine triphosphatase
Atropine, 51, 63—65, 69, 72, 87, 91
Autonomic ganglia isolated organ preparations, 9
Autonomic system, peripheral, see Peripheral autonomic system
Avoidance reactions, 74
Azaperon, 41
Azidomorphine, 14, 79

B

Behavioral effects
  conditioned behavior, 73—74
  in vivo studies, 45—53
  localized intracerebral application studies, 66—73
  neurotransmitter turnover studies, 75, 78, 88, 90—92

Benzodiazepine, 65
Benzomorphane, 36
Bezitramide, 39
Bicuculline, 65—66
Binding, receptor, see Receptor binding assay
Biochemical study methods, 16—18
Brain
    behavioral effects studies, 47—53
    catecholamines, drugs affecting, 53—63
    localized intracerebral applications studies, 66—73
    naloxone antagonism studies, 33—35
    neurotransmitter turnover studies, 75—94
    nonopiate functions of, stimulus properties and, 41—42
    serotonin in
        depletion, 41, 44—45, 53—55
        drugs affecting, 53—63
    sites of analgesic effects, 44—45
Brain peptidase preparations, 17

C

C', fragment, 17
Calcitonin-like structure, 18
Calcium ion, 37, 80, 89, 91—93
cAMP, see Cyclic AMP
Carbachol, 69
Cardiovascular system isolated organ preparations, 9
Catalepsy, 46, 48, 51—53, 64, 67—69, 71—72, 87—88
Catatonia, 51, 78
Catecholamine
    behavioral effects studies, 47—48, 50, 52—53
    conditioned behavior studies, 74
    drug interaction studies, 53—63
    localized intracerebral application studies, 71—72
    naloxone antagonism studies, 30
    neurotransmitter turnover studies, 84—88, 91
Catecholaminergic mechanisms, 47—48, 53, 71—72, 74
Cathepsin D, 17
Cations, effect on analgesia, 80
Cat nictitating membrane isolated organ preparations, 9, 14—16
Caudate nucleus
    localized intracerebral application studies, 68—69, 71
    neurotransmitter turnover studies, 79, 81, 90
Caudate-putamen, behavioral effects studies, 52
Central nervous system function, isolated organ models, 7—8
Central nucleus, amygdala, localized intracerebral application studies, 71
Centromedian, localized intracerebral application studies, 66
Cerebellum, neurotransmitter turnover studies, 90—91
Cerebral cortex, neurotransmitter turnover studies, 76, 78—79, 81

cGMP, see Cyclic GMP
p-Chloro-phenylalanine, 50, 52, 55—57, 72
Chlorpromazine
    behavioral effects studies, 48, 51—52
    drug interaction studies, 61—63
    localized intracerebral application studies, 71
    stimulus properties studies, 38, 41
Choline acetyl transferase, 79
Cholinergic agents, 51
Cholinergic mechanisms, 47, 75, 79
Cholinergic stimulants, 41
Cholinergic transmission, stimulation or inhibition of, effects of, 63—65
Cholinesterase, 77
Cholinesterase inhibitors, 63—64
Cholinolytic drugs, 63—65
Cholinomimetic drugs, 63—65
Cinanserine, 70
Clonidine, 41—42, 50, 62
Clorgyline, 55
Cocaine, 61—62
Codeine, 39, 51, 54, 56, 73, 77, 81
Competitive antagonism, naloxone, 31—33
Conditioned behavior, effects on, see also Behavioral effects, 73—74
Cross-reactivity, 6—7
Cross-tolerance, 13
Cyclic AMP, 88—91
Cyclic GMP, 89—91
Cyclic nucleotides, dopamine-sensitive, 88—91
Cycloheximide, 38
Cyclozocine
    analgesic effect studies, 43
    naloxone antagonism studies, 31, 35—36
    neurotransmitter turnover studies, 76, 84
    stimulus properties, 38—40
Cyproheptadine, 56, 62, 71

D

DDC, see Diethyldithiocarbamate
Dependence
    isolated organ studies, 7, 12—13
    naloxone antagonism studies, 36
    neurotransmitter turnover studies, 77, 78
Deprenyl, 55
Depressant effects, studies of, 46—48, 50, 52, 67—72, 91—92
Descending serotoninergic pathways
    localized intracerebral application studies, 70
    neurotransmitter turnover studies, 81—83
Descending spinal pathways, analgesic effect studies, 44
Descending spinopetal pathways, analgesic effect studies, 44
Desipramine, 58
Dexetimide, 41
Dextromoramide, 39, 89
Dextrorphan, 41, 71, 75—76, 89, 92—93
Diethyldithiocarbamate, 50, 59
3,4-Dihydroxyphenylacetic acid, 84—86
5,6-Dihydroxytryptamine, 83

Diprenorphine, 31, 34, 36—37
Disulfiram, 59
DOPA, 84
l-Dopa, 45, 51
   drug interaction studies, 53—55, 57, 60—61
Dopamine
   analgesic effects studies, 45
   drug interaction studies, 56—61, 63
   liberation, suppression of, 78
   localized intracerebral application studies, 69, 71—72
   neurotransmitter turnover studies, 78, 84—91
   turnover, 51, 84—88
Dopamine antagonists, 71
Dopamine β-hydroxylase, 50, 58—59
Dopamine receptor, 13—14, 16, 40—41, 47, 51
Dopamine receptor blockers, 41, 48, 71—72
Dopamine receptor stimulants, 51—52
Dopaminergic autoreceptors, 87—88
Dopaminergic transmission, 71, 87—88
Dopamine-sensitive cyclic neucleotides, 88—91
Double logarithmic plot, 4
Double-reciprocal plot, see Lineweaver-Burk plot
Drug interaction studies, 53—66
Dynorphin, 4
Dynorphin (1-13), 12—13, 17

E

Electric shock test
   drug interaction studies, 56, 58, 60, 62, 64
   neurotransmitter turnover studies, 80
Endogenous opioid peptide, see Opioid peptide
Endorphin, 36
α-Endorphin, 7, 17
γ-Endorphin, 17
Endorphin receptor, 47
Enkephalin, 5—8, 36, 87
   biochemical studies, 18
   discovery of, 7—8
   nerves containing, 10
   radioimmunoassay, 5—7
²Nle⁵-Enkephalin, 16
Enkephalin analogs, 9, 14—15, 31
Enkephalin congeners, 6
α-EP, radioimmunoassay, 5
β-EP
   biochemical studies, 17
   isolated organ studies, 13—16
   N-acetylation, 6
   naloxone antagonism studies, 31—32, 36
   radioimmunoassay, 5—6
Ephedrine, 62
Epinephrine, 50
Equilibrium dissociation constant, 10, 14—16
Ethanol, 41
Ethylketazocine, 50
Ethylketocyclazocine, 36, 40
Etonitazen, 39
Etorphine, 31, 43, 47, 67, 69
Excitatory effects, studies of, 46—50, 52, 67, 69—70, 78

Explosive motor behavior, 69

F

Feline mania, 48, 63
Fenfluramine, 55—56
Fentanyl, 39—42, 67, 70, 85
FK 33-824, 14
FLA-63, 47
Flinch jump test, drug interaction studies, 56
Fluphenazine, 56, 71
Food intake, reduced, causes of, 30
Frog sympathetic ganglia isolated organ preparations, 9
Funiculus dorso-lateralis, analgesic effect studies, 44

G

GABA, 58, 65—66, 87, 94
Ganglia, autonomic, see Autonomic ganglia
Gangliosides, neurotransmitter turnover studies, 93
Globus pallidus
   behavioral effects studies, 52
   localized intracerebral application studies, 67—68, 71
Guanylate cyclase, 91
Guinea pig ileum isolated organ preparations, 7—8, 10—12, 14—16

H

Half-reciprocal plot, 4
Haloperidol
   behavioral effects studies, 50, 52
   drug interaction studies, 61—63
   localized intracerebral application tests, 71
   neurotransmitter turnover studies, 87—88
   stimulus properties studies, 40—41
Heroin, 39, 50, 61, 77, 82
5-HIAA, see 5-Hydroxyindoleacetic acid
Hill plot, 4
Hippocampus, neurotransmitter turnover studies, 78—79
Histamine, 8, 94
Homogeneity, opiate receptors, 33—35
Homovanillic acid, 84—86
Hot plate test
   analgesic effect studies, 43, 50
   drug interaction studies, 53—62, 64
   naloxone antagonism studies, 31—34
HSE, see Human serum esterase
5-HT, see also Serotonin, 8, 10
   turnover, 80—84
Human serum esterase, 94
6-Hydroxydopamine, 59—60
5-Hydroxyindolacetic acid, 81, 83
5-Hydroxytryptophan, 53—57, 83—84
Hyperglycemic effects, 94

I

Imipramine, 65
Immobility, 46
Indomethacin, 41
Inflamed foot test
　analgesic effect studies, 43
　drug interaction studies, 54, 56—57, 59, 62—63
Inhibition constant, 4
Intestine, see Large intestine; Small intestine
Intracerebral application, localized, studies, 66—73
In vitro analysis, 2—18
In vivo effects, opiates, see also specific effects by name, 30—94
Ionophore X537A, 92
Iproniazid, 55
Isolated organ technique, 7—16, 31, 53, 64
　most frequently applied organs, 10—16
　peripheral autonomic and motor system preparations, 9—10
　sensitive organs, 8—9

K

16K, fragment, 18
Ketamine, 41
Ketocyclazocine, 36, 40

L

Lanthanum ion, 80, 92
Large intestine isolated organ preparations, 8
Lateral reticular nucleus, localized intracerebral application studies, 66, 68
Learning, modifiction of, 73—74
Leu-E, 10, 12, 16
Levallorphan, 10, 40, 77
Levorphanol
　behavioral effects studies, 47, 50
　conditioned behavior studies, 73—74
　localized intracerebral application studies, 67, 69, 71
　naloxone antagonism studies, 31, 35, 37
　neurotransmitter turnover studies, 75—77, 81—82, 84—85, 89—90, 92
　stimulus properties studies, 39
Ligand-receptor interactions, 2—5, 36—37
Lineweaver-Burk plot, 3—4
Lipid metabolism, 93
Liver activity, altered, 94
Localized intracerebral application studies, 66—73
Locomotor activity, changes in, 45—51
Locus coeruleus, analgesic effect sutides, 45
Longitudinal muscle, guinea pig ilea, contraction inhibitors, 8
Loperamide, 39
β-LPH
　ACTH/β-LPH precursor, 17—18
　C' fragment, 17
　isolated organ studies, 14
　radioimmunoassay, 5—7
LSD, 38, 40

M

Magnesium ion, 37, 84—92
Manganese ion, 92
Medial thalamus, analgesic effect studies, 45
Median eminence, neurotransmitter turnover studies, 86
Medulla oblongata, nucleus reticularis gigantocellularis, see Nucleus reticularis gigantocellularis
Memory, modification of, 73—74
Meperidine
　analgesic effect studies, 43
　drug interaction studies, 54—57
　localized intracerebral application studies, 68
　naloxone antagonism studies, 31
　neurotransmitter turnover studies, 75—82, 90
　stimulus properties studies, 39—40
Mescaline, 41, 61—62
Mesolimbic brain areas, behavioral effects studies, 52
D-Met, 14, 16, 31, 33, 49
Metabolic effects, 75—94
Met-E
　biochemical studies, 17
　isolated organ studies, 9—10, 12, 14—16
Met-E-Arg⁶-Phe⁷, 18
Met-E-OMe, 16
Methadone
　behavioral effects studies, 47, 51
　drug interaction studies, 56, 58
　localized intracerebral application studies, 67—69
　naloxone antagonism studies, 31, 35, 37
　neurotransmitter turnover studies, 75—78, 81—82, 84—85, 87, 89, 93
　stimulus properties studies, 39—40
3-Methoxytyramine, 86
α-Methyl DOPA, 52, 62
α-Methyltyrosine, 52, 84, 91
α-Methyl-p-tyrosine, 47, 50, 56, 58, 71, 74
Methysergide, 56, 62, 70, 83
Met/Leu-E, 13
Met/Leu-E precursor, 18
Mixed agonist-antagonists, 35, 37, 39—40, 43
Monoamine oxidase inhibitors, 55
Monoamines, 57, 60
Morphine
　analgesic effect studies, 42—45
　behavioral effects studies, 46—53
　conditioned behavior studies, 73—74
　dependence on, see Dependence
　drug interaction tests, 53—66
　isolated organ studies, 7—16
　localized intracerebral application studies, 66—73
　naloxone antagonism studies, 31—38

naloxone antagonists, pA$_2$ values, 33—38, 44, 92
neurotransmitter turnover studies, 75—94
stimulus properties, 38—42
tolerance to, see Tolerance
Morphine congeners
  in vitro studies, 10, 12—14
  in vivo studies, 35, 38, 47, 51, 53, 77—78, 88
Morphine derivatives, 75
Morphine surrogates, 2—4, 35
Motor activity, changes in, 45—52
Motor behavior, explosive, 69
Motor nerve terminal isolated organ preparations, 9—10
Motor system isolated organ preparations, 9—10
Mouse vas deferens isolated organ preparations, 12—14, 16, 36
γ-MSH, 18
Muscarinic agents, 63—64
Muscarinic-cholinergic receptors, 16
Muscimol, 65

N

Nalorphin
  behavioral effects studies, 50
  isolated organ studies, 15
  naloxone antagonism studies, 31, 34—37
  neurotransmitter turnover studies, 76—77, 81—82
  stimulus properties studies, 38—40
Naloxone
  antagonism by
    analgesic effect studies, 43—44
    competitive nature of, 31—33
    in vivo studies, 30—38
    localized intracerebral application studies, 66, 70—71
    neurotransmitter turnover studies, 75, 77, 79, 81, 83, 87, 91
  behavioral effects studies, 46—48, 50—52
  conditioned behavior studies, 74
  drug interaction studies, 64, 66
  isolated organ studies, 10, 12—13, 15—16
  localized intracerebral application studies, 66, 69—71
  neurotransmitter turnover studies, 75—79, 81—83, 86—87, 90—94
  pA$_2$ values against morphine, 33—38, 44, 92
  stimulus properties studies, 40, 42
³H-Naloxone, 11
Naltrexone
  equilibrium dissociation constant values, 14—16
  isolated organ studies, 10, 12—16
  naloxone antagonism studies, 36
Narcotic analgesics, see also Analgesic effects, 39, 43, 52
Narcotic antagonists, see Antagonists, narcotic
Narcotic drugs
  analgesic effect studies, 42—45
  antagonism by naloxone, 30—38

behavioral effects studies, 45—53
drug interaction studies, 53—66
interaction with opiate receptors, 35—37
localized intracerebral application studies, 66—73
stimulus properties studies, 38—42
α-Neo-endorphin, 17
Neostigmine, 64
Neuroleptic agents, 51—52, 61, 87
Neurotransmitter turnover studies, 75—94
Nicotine, 8
Nictitating membrane isolated organ preparations, see also Cat nictitating membrane isolated organ preparations, 9
Nonopiate functions, brain, stimulus properties and, 41—42
Noradrenaline
  analgesic effect studies, 45
  behavioral effects studies, 47
  drug interaction studies, 57—61, 63, 65
  localized intracerebral application studies, 69
  neurotransmitter turnover studies, 84—85, 87, 91
  stimulus properties studies, 41
Noradrenergic medullo-spinal pathway, neurotransmitter turnover studies, 86
Norepinephrine, 13—14, 16, 50
Normetanephrine, 86
Normorphine, 12—16
Nucleus accumbens
  drug interaction studies, 56
  localized intracerebral application studies, 68, 71—72
  neurotransmitter turnover studies, 78, 86
Nucleus amygdaloidus centralis, behavioral effects studies, 52
Nucleus caudata, see Caudate nucleus
Neuclues raphe dorsalis, neurotransmitter turnover studies, 78, 83
Nucleus raphe magnus, see also Raphe complex
  analgesic effect studies, 44
  neurotransmitter turnover studies, 86
  stimulus properties studies, 42
Nucleus reticularis gigantocellularis
  localized intracerebral application studies, 68, 70
  neurotransmitter turnover studies, 85—86
Nucleus reticularis (para) gigantocellularis, localized intracerebral application studies, 66, 69—70
Nucleus septi, neurotransmitter turnover studies, 78

O

6-OHDA, see 6-Hydroxydopamine
Olfactory tubercle, neurotransmitter turnover studies, 86
  operant behavior, effects on, 74
Opiate agonists and antagonists, see Agonist, opiate; Antagonist, opiate
Opiate receptor, see Receptor, opiate

Opiates
  in vitro studies, 2—18
  in vivo studies, 30—94
Oxotremorine, 64, 77
Oxylorphan, 40
Oxymorphone, 15, 39
Oxypertine, 54, 58

P

pA_2
  derivation of, 10
  determination in vivo, 31—33
  naloxone, against morphine, 33—38, 44, 92
  narcotic antagonists of morphine, 33—38
Para-chloro-phenylalanine, see p-Chloro-phenylalanine
Parafascicular nuclei of thalamus, localized intracerebral application studies, 66, 68
Pargyline, 55
PCPA, see p-Chloro-phenylalanine
Pentazocine
  analgesic effect studies, 43
  naloxone antagonism studies, 31, 35—37
  neurotransmitter turnover studies, 76, 81—82, 84—85, 90
  stimulus properties, 40
Pentobarbitone, 38, 41, 65
Pentolinium, 91
Periaqueductal gray matter
  analgesic effect studies, 44—45
  behavioral effect studies, 47
  conditioned behavior studies, 73
  localized intracerebral application studies, 66—72
  neurotransmitter turnover studies, 82—83, 86, 92, 94
  stimulus property studies, 41
Peripheral autonomic system isolated organ preparations, 9—10
Periventricular area, localized intracerebral application studies, 66—68
Perphenazine, 62
Pethidine, 51
PGE, see Prostaglandin E
Pharmacological effects, opiates, 2—3, 30—95
  receptors mediating, 30—95
Pharmacological manipulation, opiate receptor affinity for agonists and antagonists affected by, 37—38
Phenacetive, 41
Phenazocine, 39, 77, 84
Phenelzine, 55
Phenoxybenzamine, 47—48, 62—63
Phentolamine, 62—63, 69—70, 86, 91
Phenylbutazone, 41
Phosphatidylinositol, 93
Phosphatidylserine, 93
Phosphodiesterase, 80, 89—91
Phosphoinositides, 93
Phospholipid, 93
Phosphorus, 93

Physiological conditions, opiate receptor affinity for agonists and antagonists affected by, 37—38
Physostigmine, 41, 50, 64, 82
Pilocarpine, 41, 64
Pimozide, 47, 52, 62, 69, 71
Piperoxane, 71
Posterior hypothalamus, localized intracerebral application studies, 66—67
Potassium ion, 80, 89—92
Practolol, 62
Preoptic forebrain region, analgesic effect studies, 45
Presynaptic α-adrenergic receptor stimulants, 41
Pro-EA, 14, 16, 31, 33, 49
Profadol, 57
Prolactin, 86
Propanolol, 48, 62—63, 69, 91
Proportion graph, 4
Propoxyphene, 56, 58, 77
Prostaglandin, 94
Prostaglandin E_1, 16, 66, 88
Prostaglandin E_2, 11
Protein synthesis, 84, 93—94
Psychomimetic drugs, 40—41
Psychostimulant drugs, 61
Psychotropic drugs, 42
Pure narcotics, analgesic effects, 43
Purinergic nerves, 10—11

R

Rabbit ear artery isolated organ preparations, 9, 16
Radioimmunoassay, 5—7
Raphe complex
  analgesic effect studies, 44—45
  behavioral effects studies, 52—53
  localized intracerebral application studies, 66, 69—70
  neurotransmitter turnover studies, 78, 81, 83, 86
  stimulus properties studies, 42
Rat vas deferens isolated organ preparations, 9, 14, 36
Receptor, opiate
  affinity for agonists and antagonists, changes in, 37—38
  behavioral effect studies, 45—53
  binding studies, 2—5
  homogeneity, 33—35
  in vivo functions, 30—95
  localized intracerebral application studies, 66—73
  narcotic drugs interacting with, 35—37
  neurotransmitter turnover studies, 78—79, 86—87, 90—94
  pharmacological effects mediated by, 30—95
α-Receptor, opiate, 4, 36
δ-Receptor, opiate, 4, 12—13, 36
ε-Receptor, 14, 36
$\varkappa$-Receptor, opiate, 4

μ-Receptor, opiate, 4, 12—13, 36, 50
χ-Receptor, opiate, 4, 12—13, 36, 50
Receptor binding assay, 2—5, 36—37
Receptor blocking and stimulating agents, 61—63
Reserpine, 48, 50, 53—55, 63, 65, 77, 81
Reticular formation, localized intracerebral application studies, 69
Ribonucleic acid synthesis, 93—94
Rigidity, 46, 51—52
RNA, see Ribonucleic acid
Running fit, 78, 50

S

Scatchard plot, 4
Scopolamine, 41, 51, 64—65
Sedato-hypnotic drugs, 41
Septum, neurotransmitter turnover studies, 79
Serotonin
    behavioral effects studies, 47, 50, 52—53
    depletion, brain, effects of, 41, 44—45, 53—55
    drug interaction studies, 53—63, 65
    isolated organ studies, 8, 12
    localized intracerebral application studies, 70—72
    turnover, 80—84
Serotoninergic pathways
    ascending, neurotransmitter turnover studies, 81—82
    descending
        localized intracerebral application studies, 70
        neurotransmitter turnover studies, 81—83
Serotonin receptor blockers, 55—56, 70
Serotonin receptor stimulators, 55, 70
Serum glutamate-oxalate transaminase, 94
Serum glutamate-pyruvate transaminase, 94
SGOT, see Serum glutamate-oxalate transaminase
SGPT, see Serum glutamate-pyruvate transaminase
Shock titration test
    analgesic effect studies, 43
    drug interaction studies, 64
    naloxone antagonism studies, 33—34
Small intestine isolated organ preparations, 8
Sodium ion, 37, 80, 89—92, 94
SP, 11
Spinal cord, neurotransmitter turnover studies, 83, 86
Spontaneous motor activity, changes in, 45—52
Stereotypy, 52—53, 87
Stimulus property studies, 38—42
    animal, 39—41
    human, 38—39
    suppression by modification of nonopiate functions of brain, 41—42
Straub phenomenon, 45—46, 51
Striatum
    behavioral effects studies, 51—52
    drug interaction studies, 60
    localized intracerebral application studies, 68, 70—72
    neurotransmitter turnover studies, 78—79, 86—90
Substance P, 10, 12, 14, 42
Substantia nigra
    drug interaction studies, 60
    localized intracerebral application studies, 66, 68, 70—71
    neurotransmitter turnover studies, 78, 88
Sufentanyl, 41
Sulpiride, 16

T

Tail compression test
    drug interaction studies, 53—65
    naloxone antagonism studies, 31, 33—34
Tail erection, see Straub phenomenon
Tail flick test
    analgesic effect studies, 42—43
    behavioral effects studies, 73
    drug interaction studies, 52—65
    naloxone antagonism studies, 31, 33—34
    neurotransmitter turnover studies, 80, 83
Tail immersion test
    analgesic effect studies, 42—43
    drug interaction studies, 57—58
Tail pinch test, drug interaction studies, 54, 57, 59, 62, 64—65
TAT, see Tyrosine aminotransferase
Tetrabenazine, 48, 63
Thebaine, 40
Thiamin, 94
Tolerance
    behavioral effects studies, 46—47, 50
    conditioned behavior studies, 74
    cross-, 13
    isolated organ studies, 7, 12—13
    naloxone antagonism studies, 36—38
    neurotransmitter turnover studies, 75, 88, 92—94
Tolmetin, 41
Tranylcypromine, 55
Tryptophan, 83—84
Tryptophan hydroxylase, 83
Turning behavior, morphine-induced, 69
Tyrosine, 57, 84, 93
Tyrosine aminotransferase, 94
Tyrosine hydroxylase, 84

U

U-14624, 47
Uridine, 93
Urogenital tract isolated organ preparations, 9

V

Vas deferens, see Mouse vas deferens; Rat vas deferens

Ventral nucleus, thalamus, localized intracerebral application studies, 71
Ventral thalamus, localized intracerebral application studies, 67—68, 72
Ventricle, localized intracerebral application studies, 66—67
VIP, 10—11
Vocalization after-discharge test
　analgesic effect studies, 43
　drug interaction studies, 56, 59, 62, 64
Vocalization response test, analgesic effect studies, 42—43

W

Weight loss, causes of, 30

Withdrawal symptoms, 12, 77
Writhing test
　analgesic effect studies, 43
　drug interaction studies, 56, 62, 64—65
　naloxone antagonism studies, 31, 33—34

X

X537A ionophore, 92

Y

Yohimbine, 62

JULIA TUTWILER LIBRARY
LIVINGSTON UNIVERSITY
DATE BOOK IS DUE BACK

This book is due back on or before the last date stamped in back. If it is returned after that date, the borrower will be charged a daily fine. In checking the book out the borrower assumes responsibility for returning it.

NO "OVER-DUE" NOTICE WILL BE SENT OUT FROM THE LIBRARY.